SPACE, PLACE AND MENTAL HEALTH

Geographies of Health

Series Editors
Allison Williams, Associate Professor, School of Geography and Earth
Sciences, McMaster University, Canada
Susan Elliott, Dean of the Faculty of Social Sciences,
McMaster University, Canada

There is growing interest in the geographies of health and a continued interest in what has more traditionally been labeled medical geography. The traditional focus of 'medical geography' on areas such as disease ecology, health service provision and disease mapping (all of which continue to reflect a mainly quantitative approach to inquiry) has evolved to a focus on a broader, theoretically informed epistemology of health geographies in an expanded international reach. As a result, we now find this subdiscipline characterized by a strongly theoretically-informed research agenda, embracing a range of methods (quantitative; qualitative and the integration of the two) of inquiry concerned with questions of: risk; representation and meaning; inequality and power; culture and difference, among others. Health mapping and modeling has simultaneously been strengthened by the technical advances made in multilevel modeling, advanced spatial analytic methods and GIS, while further engaging in questions related to health inequalities, population health and environmental degradation.

This series publishes superior quality research monographs and edited collections representing contemporary applications in the field; this encompasses original research as well as advances in methods, techniques and theories. The *Geographies of Health* series will capture the interest of a broad body of scholars, within the social sciences, the health sciences and beyond.

Also in the series

Healing Waters
Therapeutic Landscapes in Historic and Contemporary Ireland
Ronan Foley
ISBN 978 0 7546 7652 2

Towards Enabling Geographies:
'Disabled' Bodies and Minds in Society and Space
Edited by Vera Chouinard, Edward Hall, and Robert Wilton
ISBN 978 0 7546 7561 7

Geographies of Obesity: Environmental Understandings of the Obesity Epidemic
Edited by Jamie Pearce and Karen Witten
ISBN 978 0 7546 7619 5

Space, Place and Mental Health

SARAH CURTIS
University of Durham, UK

ASHGATE

© Sarah Curtis 2010

Published by
Ashgate Publishing Limited
Wey Court East
Union Road
Farnham
Surrey, GU9 7PT
England

Ashgate Publishing Company
Suite 420
101 Cherry Street
Burlington
VT 05401-4405
USA

www.ashgate.com

British Library Cataloguing in Publication Data
Curtis, Sarah, 1954-
   Space, place and mental health. -- (Ashgate's geographies
   of health series)
   1. Mental health--Environmental aspects. 2. Mental
   illness--Environmental aspects. 3. Environmental
   psychology. 4. Mind and body. 5. Medical geography.
   I. Title II. Series
   616.8'9071-dc22

Library of Congress Cataloging-in-Publication Data
Curtis, Sarah.
   Space, place and mental health / by Sarah Curtis.
      p. cm. -- (Ashgate's geographies of health series)
   Includes bibliographical references and index.
   ISBN 978-0-7546-7331-6 (hardback) -- ISBN 978-0-7546-9040-5 (ebook)
1. Medical geography. 2. Human geography. 3. Mental health. 4. World health. I. Title.

RA792.C87 2010
614.4'2--dc22

2010003964

ISBN 978 0 7546 7331 6 (hbk)
ISBN 978 0 7546 9040 5 (ebk)

Mixed Sources
Product group from well-managed
forests and other controlled sources
www.fsc.org Cert no. SA-COC-1565
© 1996 Forest Stewardship Council

Printed and bound in Great Britain by
MPG Books Group, UK

Contents

List of Figures

List of Boxes

Foreword

This book aims to provide an overview of research on mental health from a geographical perspective and to show how and why space and place are important for mental health of individuals and populations. One objective is to move forward the research agenda in this field by considering in combination fields of research which are often treated rather separately. The book aims to interest an interdisciplinary audience of researchers concerned with mental health of individuals and population. The book is also intended to offer an introduction for higher level undergraduate and postgraduate students who may be undertaking courses in health geography, public health, sociology or anthropology of health and illness. For this reason I have attempted to provide some explanation of basic concepts throughout the discussion, and some specific illustrations of research ideas and methods have been presented in 'boxes' which are intended to provide suitable cases for discussion in class. A summary of some possible learning objectives is offered at the end of each chapter (though these are illustrative rather than exhaustive) and some suggestions for introductory reading are also listed.

Acknowledgements

I would like to thank those who have helped in the production of this book and in the research on which this book is based. Special thanks go to staff at Ashgate and to Kimberley Armstrong and Jenny Laws at Durham for all their editorial assistance. Also I want particularly to thank my husband Brian Blundell for his unfailing support and good humour, without which this would not have got done.

Chapter 1

Mens Sana in Corpore Sano?
Introduction to Geographical Perspectives on
Health of the Mind and the Body

Summary

This book takes a geographical perspective on the growing public interest in mental health, happiness, and psychological 'wellbeing'. It focuses on the significance of physical and social environments for these aspects of health. The aim is to explore how and why space and place are important for mental health and psychological wellbeing. The book reviews research that investigates geographical factors associated with risk of psychological distress or mental illness, or, conversely, with the chances of enjoying a 'healthy' mental state and positive sense of wellbeing.

This chapter provides a preliminary overview of geographical perspectives, which draw on a range of theories and methods, to interpret the complex relationships between our mental health and the social and physical environment in which we live.

This chapter starts with a discussion of the scope of health geography. It introduces some themes that are important for what follows later in this book. These include geographical theories about *space*, and *place* as they relate to health geography. Also introduced in this chapter are related ideas about ways of conceptualising *risk*. Consideration is given to aspects of classic and more recent debates concerning *dichotomies of mind and body* and the *connections between* 'mental' *and* 'physical' *health*. This book focuses on selected aspects of the 'mind/body/environment complex', and the ways that 'states of mind' are described in terms of positive 'mental health' or 'psychological wellbeing', as well as aspects of 'mental illness' and *disorder*. This chapter summarises some of the different ways that researchers define, describe and measure 'mental health' as it is construed in this book. Ideas from geographies of health are given most attention, but the perspective is interdisciplinary, aiming to illustrate how geography connects with ideas from, for example, sociology, anthropology, psychology and psychiatry to help us understand spaces of risk for public mental health.

Mental health and wellbeing comprise very significant aspects of human health. The importance of the burden of mental illnesses in human populations is considered, as well as the growing preoccupation in contemporary societies with the ways that our environment relates to wellbeing, 'happiness' and quality of life.

Introducing a perspective on space, place and mental health

How and why does the social and physical environment matter for mental health and psychological wellbeing in human populations? In this book this question is examined from a number of different points of view but with a particular emphasis on ideas and research from health geography. This book aims to show that health geography makes a distinctive contribution to knowledge about mental health through its focus on the interactions between people and their social and physical environment and in the ways that it conceptualises space and place.

Although the emphasis in this book is on a geographical perspective, the aim is also to demonstrate the connections between health geography and other disciplines particularly concerned with 'states of mind' and with mental health, including philosophy, psychiatry, psychology and public health. Therefore, this first chapter summarises selected parts of the relevant literature from these various disciplines concerning 'mind-body dualism' to contextualise mental health geography. A geographical perspective on mental health also fits well with the 'social model' used in public health to conceptualise the 'wider determinants' of health, which is considered below. Socio-geographical perspectives link to the sociology of health and illness in the way that they give attention to the idea of health as *socially* as well as *medically* constructed.

The book aims to review a rich body of theory and research to show how attributes of 'space' and 'place' are associated with human mental wellbeing or with mental illness. It will review concepts of space and place that help us to understand these associations. Some aspects of physical and social environments are associated with better chances of enjoying a 'healthy' mental state and positive sense of wellbeing. Conversely, some geographical factors may be associated with a higher probability of psychological distress or mental illness for individuals or populations. In the language of epidemiology and public health, these varying chances of mental health for different groups of people are often discussed in terms of 'risks' for mental public health. Much of the discussion in this book therefore aims to review what we learn from a geographical perspective about the connections between *space, place* and *risk* for mental health. This book also considers the ways that understandings of risk are constructed socially as well as scientifically. Thus some of the literature from social epidemiology and sociology of risk is relevant to the following discussion.

Arguments explaining the detailed reasons for these links between geographical factors and mental health are developed in later chapters in this book, but in this chapter some of the main themes are introduced. The sections of this chapter deal with the following topics: the scope of health geography; geographical interpretations of 'relational' concepts of space and place; concepts of risk; theories about 'states of mind'; how far the mind and the body can be differentiated in research on health and geographical experience. Consideration is given to methods to assess mental health and wellbeing and what these tell us about the burden of mental illness and the importance of psychological wellbeing for human health.

Also discussed is some of the international evidence for the significance of mental illness and wellbeing for human health more generally.

The scope of 'health geography': perspectives on the importance of place and space for health

This book is written primarily from a health geography perspective, and it makes sense at the outset to clarify what this approach involves. Geography considers the significance for physical and mental health of *interactions between people and their environment*. It investigates why *space* and *place* are important for health variation in human populations. Contrary to a view of geography which is apparently held by some of those who are not specialist in the field, health geography it is *not* solely about cartography (i.e. mapping health variation among geographical areas). As will be shown in this book, cartography is one of many tools that are useful for geographic research on health, but a large proportion of publications in health geography do not contain any maps at all. Also health geography is an *investigative*, rather than a purely descriptive endeavour. Although *describing* the geographical patterns of variation in mental health, health care facilities and resources may form part of geographical research, a more important element comprises *explanatory* studies which aim to investigate the *reasons* for these geographically variable distributions.

This book is one of several that have reviewed aspects of health geography, so that writing from a health geography standpoint, I am able to draw a number of earlier overviews. A summary in Box 1.1 discusses key reviews of health geography which may be of interest for further reading. From this we can see that health geography has a long pedigree; authors such as Barrett (2000) have compiled scholarly collections of material relating to the subject which date back at least as far as ancient Greek writings. Contemporary health geography has grown out of the field known as *medical geography* (which is particularly concerned with geographical aspects of medically defined diseases and medical care). The term *health geography* has been introduced in recognition of the widening of geographical concern to include aspects of health and wellbeing that are not medically defined and might be considered more broadly the domain of public health (e.g. for a recent discussion of the idea of wellbeing from a health geography perspective see Fleuret and Atkinson, 2007).

Some of the authors listed in Box 1.1 give specific attention to mental health; for example, Jones and Moon (1987), and Curtis (2004), do so within overviews of geographies of health in general. A significant geographical literature has also been more specifically concerned with mental health and psychiatric care (see Box 1.2 and especially Parr, 2008). These texts are all discussed in more detail in later chapters of this book and the notes in Box 1.2 are intended only as a brief guide to further reading in the form of key texts which have contributed to the development of geographies of mental health. These examples show that health geography has been concerned with various aspects of mental health and illness including: geographical patterns of mental illness at the population level; the distribution of psychiatric services and varying access to mental health care; the types of places where psychiatric care is provided; and the geographical experience of mental illness, psychological wellbeing and learning disability.

Box 1.1   Suggestions for further reading about health geography

Readers who would like to have a more general sense of the field of health geography as a whole may be interested to read more widely in the subject. A good place to start is with a number of important overviews of the subject, several of which consider examples concerning mental as well as physical health. They include, for example, Barrett's impressive compilations of material tracing the roots of health geography back to ancient Greek writings (Barrett, 2000). The medical geography perspective has also been very thoroughly presented by Learmonth (1988) and more recently by Meade and Earickson (2000) with the arguments illustrated using many international examples of research on specific diseases. Reviews by Shannon and Dever (1974), Haynes and Bentham (1979), Joseph and Phillips (1984) have been influential in the development of geographical work on the organisation of health care, access to services and territorial justice in health care. Other major landmarks in the general literature on health geography include a book by Jones and Moon (1987) which was especially influential because of the way that it interpreted social as well as biomedical theories from a geographical perspective. Curtis and Taket (1996) also focused more particularly on the significance of social processes for geographies of health and elaborated on the connection between geographical perspectives and public health policy, and Chapter 2 of their book provides an overview of developments in health geography in recent decades. Gatrell (2002) and Curtis (2004) have also provided more up to date reviews of a range of perspectives in health geography and Gatrell's text is especially recommended for those interested to learn more about the physical environmental processes as well as the social processes that impact on human health. The field has now developed to such a level internationally that a compendium of health geography is currently in preparation (Brown and Moon, 2009) and the international strength of interest in the subject is reflected in reviews produced in languages other than English (e.g. Fleuret and Thouez, 2007).

Box 1.2   Suggestions for complementary reading on the development of geographies of mental health

A very helpful, succinct review of the evolution of literature in geographies of mental health is provided by Wolch and Philo (2000) (see also Philo and Woch, 2001; Philo, 2005; Philo 1997) summarising progress in geographies of mental health spanning the period 1970 to the present day and arguing that the agenda has shifted from studies of the locations in which psychiatric care is provided and the geographical location of people with mental illness towards research which is more focused on the role of space and place in differentiation and marginalisation of people with mental illness on the part of wider society (e.g. Parr, 2008). Research taking a longer historical perspective, has produced a series of studies of asylum and post-asylum geographies (some of which are referred to in Chapter 7 of this book. Philo (2004) provides a particularly authoritative account of the history of institutional provision for people with mental illness. Important contributions to early work in geography of mental health were also made by Smith and Giggs and Dear (e.g. See Giggs, 1973, 1975; Smith 1978; Smith and Giggs, 1988; Smith and Hanham, 1981a, 1981b; Dear and Taylor 1982) who explored the geographical distribution in the city of people with mental illness and of services provided for them, as well as the social processes associated with these distributions. Also often cited in the literature from this period is an important critique of deinstitutionalisation of psychiatric care in North America, based on a clearly formulated conceptual framework, by Dear and Wolch (1987). Moon and colleagues have provided analyses and reviews of geographies of psychiatric morbidity and care and the social construction of risk associated with mental illness (e.g. Jones and Moon, 1987; Moon, 2000; Duncan, Jones and Moon, 1995). Research using relatively advanced statistical techniques to extend our understanding of the geographical patterning of psychiatric illnesses also include studies by Congdon et al. (1996, 2006), Weich and Lewis (1998), Wiech et al. (2001), Kessler et al. (2004) and Middleton, Sterne, Gunnell et al. (2008). Also strongly recommended in the literature on mental health geographies is work by Parr (e.g. 1997, 1999, 2000, 2008) which has elucidated the experiences of people with mental illness from a geographical point of view. The collection of papers for which she was an editor (Butler and Parr, 1999) are valuable early examples of the geographical concern with embodiment and 'states of mind'. Geographers also use similar approaches to consider aspects of learning disability and impairment that would not be defined as medical conditions; for example, a collection of papers on this topic is introduced by Philo and Metzel (2005).

Theories of space and place in health geography

Health geography works with a range of theoretical and empirical approaches in which ideas about place and space are fundamental. Place and space are not seen merely as 'containers' for 'epidemiological processes' (causal pathways producing health variation). Rather, space and place *contribute* to processes helping to *constitute* health variation among individuals and populations (e.g. see Kearns, 1993; Kearns and Joseph, 1993). Consideration of these relationships has led to approaches in empirical research on population health which aim to investigate how 'compositional' attributes of the individuals making up the population interact with the 'contextual' aspects of a socio-geographical space or 'setting' that may impinge on many of the individuals embraced within it (e.g. reviewed by Cummins et al., 2007). Moreover, as I hope to demonstrate in this book, places and health have mutual effects on each other, in that states of mind and mental illness can also influence the ways that places are identified and interpreted. In earlier publications in health geography (Curtis and Rees-Jones, 1998; Curtis, 2004) key ideas in health geography were framed using the metaphor of different 'types' of 'landscape', each representing a body of theory about the connections between place and health, summarised in Figure 1.1. These relationships have been reviewed at length in my earlier book (Curtis, 2004) so I will not repeat here the detail of that discussion. In brief, we see from Figure 1.1 that health geography has viewed health inequality in human populations from a range of theoretical perspectives about space and place. These variously interpret humanistic and psychological theories about the following processes:

- our emotional and psychological response to places and their significance for sense of identity, wellbeing and identity;
- the role of space and place in processes of social and political control and resistance in the relations between dominant and subordinated social groups;
- the links between socially defined spaces in which we live (our 'habitus') and our choices and expression of taste through our private consumption of commodities and individual lifestyle practices;
- processes of collective consumption; action by which societies try to ensure socio-economic reproduction at the community and national level and promote social and territorial justice;
- epidemiological and ecological processes linking physical and bio-chemical conditions in the environment to human health.

Other conceptual frameworks have also been proposed in geography to express the various dimensions of the environment that are relevant for human health. Human disease ecology (e.g. discussed by Meade and Earickson, 2002) also conceptualises interactions between physical environmental processes, human biological attributes and human behaviours. Furthermore, from a humanist perspective in social

Landscape type	Theoretical framework
'Therapeutic' landscapes of wellbeing	Theories of sense of place/identity and emotional responses to settings
Landscapes of power and resistance	Theories of social and political control and resistance in socially defined spaces
Landscapes of material poverty	Theories of socio-economic production and structuration and geographically uneven development
Landscapes of private consumption	Theories of consumption, including private consumption, taste and lifestyle and commodification
Landscapes of collective consumption	Theories of socio-economic reproduction and social justice allied to collective action to provide goods and services
Landscapes of biological and physical hazard	Epidemiological and ecological theories concerning physical/chemical processes in the environment

Figure 1.1 Landscapes of health and risk in health geography

Source: Adapted from Curtis, 2004, p. 23.

geography, the idea of therapeutic landscapes was proposed by Gesler (e.g. Gesler 1992, 2003, 2005) and has been developed in subsequent research to explain the ways that certain landscapes offer potential for healing and promotion of good health (e.g. see edited collections by Williams, 1998, 2007). The 'therapeutic landscapes' perspective draws on several different areas of theory and is discussed in greater detail later in this book. It has proved very attractive to many researchers as a conceptual framework to organise ideas about how people experience landscape in ways that are important for their health. Although there are some variations in the way that the conceptual framework has been presented, it broadly proposes that geography should consider aspects of the physical (natural and built) environment, the social environment (comprising social relationships) and the symbolic environment (understood through the meanings attached to geographical settings).

The thinking reflected in all of these different conceptual schemes is summarised in Figure 1.2, which expresses the idea of geographical environments comprising material, social and symbolic dimensions, which merge in particular ways for every individual as they move along their lifecourse trajectory through space and time. The interactions between the person and these changing settings is important for health outcomes of the individual, and the individual also contributes to the health environment. This model implies a complex set of relationships which require a 'relational' view of health geography, as discussed below. This perspective can best be considered as an organising framework to integrate different fields of theory, rather than as a single theory. The more detailed theoretical arguments that underpin the framework are discussed in later chapters of this book.

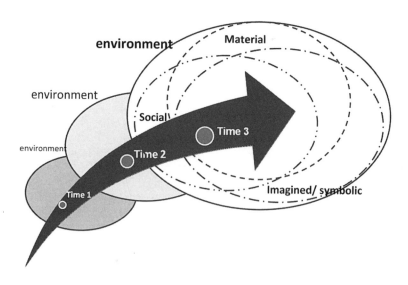

Figure 1.2   Diagram representing a person's space/time trajectory through changing environments relating to human health

Conceptualisations of complex, relational spaces

Cummins, Curtis, Diez-Roux and Macintyre (2007) have recently reviewed trends in health geography which reflect recent developments in geographical thinking about space and place. These perspectives draw especially on the interdisciplinary literature concerning two conceptual frameworks:

- *actor network theories* were developed for example from writings by Latour (1996) and discussed by geographers such as: Murdoch (1998), and Murdoch, Marsden and Banks (2000). Actor Network theory (ANT) focuses especially on relationships among the various elements and spaces that make up social and physical environment;
- theories about *complexity* concern the ways that human-environments develop and evolve in time and space and the ways that complex systems operate at different geographical scales. These theories were formulated by a number of different authors and are well reviewed, for example, from an interdisciplinary perspective by Waldrop (1994); and from a social science perspective by Byrne (1998) and Curtis and Riva (2010a, 2009b).

Key aspects of the 'relational' view of space are summarised in the right hand column of Figure 1.3 from Cummins, Curtis, Diez-Roux and Macintyre (2007) which draws on writing by a number of geographers including: Massey (1999, 2005); Castree (2004); Conradson (2005); Graham and Healey (1999); Hudson (2004);

'Conventional' view	'Relational' view
spaces with geographical boundaries drawn at a specific scale	nodes in networks, multi-scale
separated by physical distance	separated by socio-relational distance
resident local communities	populations of individuals who are mobile daily and over their lifecourse
services described in terms of fixed locations often providing for territorial jurisdictions, distance decay models describe varying utility in space'	'layers' of assets available to populations via varying paths in time and space. Euclidean distance may not be relevant to utility.
area definitions relatively static and fixed	area definitions relatively dynamic and fluid
characteristics at fixed time points, e.g. 'deprived' versus 'affluent'	dynamic, 'emergent' characteristics e.g. 'declining' versus 'advancing'
culturally neutral territorial divisions, infrastructure and services	territorial divisions, services and infrastructure imbued with social power relations and cultural meaning
contextual features described systematically and consistently by different individuals and groups	contextual features described variably by different individuals and groups

Figure 1.3   'Conventional' and 'relational' understandings of 'place'

Source: Steven Cummins, Sarah Curtis, Ana V. Diez-Roux and Sally Macintyre (2007) Understanding and representing 'place' in health research: a relational approach, *Social Science and Medicine* 65(9): 1825–38. Reprinted from *The Lancet* by permission of Elsevier.

Yeung (2005), who have discussed the significance of these relational perspectives and argued that they demand that we reappraise and revise 'conventional' ways of thinking about space and place (which are also summarised in the left hand column of Figure 1.3).

The 'conventional' view of space has considered geographical areas defined in terms of areas (places) that can be delineated by geographical boundaries (which might be 'natural' such as coasts, rivers, mountain ranges) or human constructs such as territorial boundaries, roads, railways and other linear features in the landscape. These places, or 'parcels of space' are positioned relative to each other in Euclidean space. The separation between places is represented in terms of Euclidean spatial distance (straight line distance 'on the ground' or perhaps travel distance along winding routes). This geographical view of space divided up into areas is often represented as being neutral from a socio-cultural point of view and universally understood (although political/territorial boundaries can be the source of conflict and dissent). The places comprised of these conventional spaces are often characterised in terms of local 'communities', made up of the populations with fixed places of residence inhabiting the space and of the human activities and social networks that typically operate within these areas and contribute to their character. Thus we tend to talk about areas that are 'deprived' or 'affluent', 'urbanised' or 'rural', 'industrial' or 'agricultural', socially 'close knit' or 'fragmented'. The geography of services such as health care are also viewed in terms of distinct, contiguous, and often rather local catchment areas with identifiable geographical boundaries to which services are usually provided from bases in fixed facilities such as hospitals or clinics. These definitions and attributes of areas are generally represented as though they were static, stable features of these areas over time (though people may change their place of residence, and boundary definitions and the characteristics of areas within boundaries may change from time to time).

Relational perspectives put in question most of the assumptions outlined above. According to relational interpretations, space is more usefully thought about in terms of a complex set of nodes in networks, which are not organised in a static way according to Euclidean distance but much more fluid and dynamic and subject to multiple interpretations by different people and agencies at any one time. Processes of globalisation and 'space time compression' mean that spatial distance on the ground is less important for connectedness than the time taken to communicate from one place to another (which in terms of satellite telephonic and electronic internet communication, for example, can be negligible over very long distances whereas communication between close, but 'poorly connected' places may be relatively slow). Socially constructed definitions of space are crucial, so that places that are close together in Euclidean space may also be very distant in terms of the social relations between the populations living in them. Socially meaningful 'communities' may be constituted between individuals who each occupy multiple spaces or groups who spend most of their time widely dispersed in spatially distant locations. Highly mobile, 'globe trotting' executives, or poor migrants moving to set up 'home places' in new countries, while maintaining links to their extended families in their countries of origin, exemplify these tendencies. Ideas about the character and attributes of places, being socially constructed, may be simultaneously viewed quite differently by different social and cultural groups of people. As discussed in Chapters 4 and 5, characteristics of places are constructed 'from within' according to views and attributes of local inhabitants, but also 'from without' by perceptions of people who have never physically entered the spaces in question, but form views based on representations of the place as 'healthy' or 'unhealthy' in media and other public discourses. These socially constructed attributes of places are therefore not socially and politically neutral or universal but tend to reflect social relationships between more and less powerful groups in society. The local activities going on in places and their socio-economic functions (such as employment patterns, local housing markets) are not entirely dependent on local conditions but strongly affected by national and world scale processes operating in global circuits of economic and cultural capital. This means that we need to conceptualise processes as simultaneously local and global, rather than operating primarily at one scale or another.

Furthermore, relational space needs to be seen in a temporal perspective, as dynamic and fluctuating, involving individuals constantly moving through space on different trajectories, and spaces constantly evolving in their relationship to other parts of the spatial network. For example, Massey (2005: 9) describes space in dynamic terms as "always in the process of being made. It is never finished, never closed". This vision of space is therefore fluid and liable to be defined and redefined in variable ways over time, or even at any one moment depending on the particular viewpoint or perception of the observer. Complexity theory also involves the idea of 'emergence' whereby particular conjunctions in space and time of myriad, small, and sometimes relatively simple, interactions constantly generate potential instability in natural and human systems. Such systems are constantly evolving and the pace of change may vary between relatively stable, ordered phases to a condition expressed in the title of Waldrop's (1994) discussion as the 'edge of chaos'. At the 'edge of chaos' a new pattern of organisation and 'order' may emerge and become provisionally established, or systems may 'dissolve' into periods that appear very chaotic and disorganised. One feature of emergence is that systems exhibit 'path dependency', having a 'history' which partly influences how they may evolve in future. This underlines that the 'lifecourse' experience of systems, places and people, across time and space, is important to their instantaneous condition in any one time and space.

These aspects of human experience of space and place are not new phenomena (for instance, people have moved around the globe and mixed together in fairly large and complex communities for centuries). However, as 'globalisation' becomes an increasingly dominant feature of life for many people around the world, and cities become increasingly large and complex, these aspects of relational space seem increasingly significant and important for health geography. Furthermore the research methods and tools we have available to describe and analyse processes and phenomena in relational space have considerably advanced. For example, more powerful computers and more advanced and sophisticated computer software now allow us to model the interactions between attributes of individual people and context as they affect individual outcomes and system changes over time. We have enhanced ability to track people and other mobile features of the environment over time and space using Global Positioning Systems (GPS) and interpret these data using more sophisticated analytical Geographical Information Systems (GIS) (e.g. reviewed by Kwan, 2004). Also, geography, like other social sciences is rapidly expanding and refining its use of intensive research methods that allow individuals to express their perceptions and feelings in a variety of different ways including photo-voice, drama, and participative techniques (Pain, 2004; Pain and Kindon, 2007).

Cummins, Curtis, Diez-Roux and Macintyre (2007) draw attention to new research agendas that have emerged in health geography as a result of these ideas about relational space and advances in research methods. These include an increasing emphasis on exploring the *interactions* between individual people and the context in which they live their lives (which moves beyond concerns in the 1990s to establish that contextual as well as individual attributes had *independent* effects on health). Ideas from relational geography raise questions over whether it is really possible, in studies of groups of individual people in different geographical contexts, to distinguish individual and contextual attributes as independent dimensions of processes relating to health differences. The most fruitful way forward is now to focus on the health implications of the mutual connections between people and their environment.

Geographers are also building into their research more sophisticated ways of considering temporal as well as spatial dimensions and a more complex approach to scale. There is evidently scope to further expand research which considers processes relating to human health in time as well as space, with growth of potential for more studies of human lifecourse experience and of trajectories of various phenomena and processes in social and physical environments. We need to do more to develop understanding about processes operating on human health across a continuum of spatial and temporal scales, from the global to the very local and encompassing long term as well as short term trends. Because we now understand these processes in terms of complex, networked systems there is also considerable scope to do more work around the interactions between social and physical aspects of environment and between human activity and 'natural' phenomena. Potential for work on these themes is illustrated, for example, by recent research initiatives in many countries to consider how human health fits into complex processes such as climate change and globalisation of economies and cultures.

Conceptualisation of risk and 'risk society'

Connected with these discussions about the relational characteristics of space and the social construction of our understanding of the world we live in, are concepts of *risk*. Risk, as interpreted by researchers in many disciplines, including geography, is increasingly seen as a socially, as well as scientifically constructed idea, which needs to be viewed in light of the notions introduced above of relational space, scale and complexity of the world we live in.

Influential arguments about risk in society are articulated in widely cited publications by Beck (1992 and 1999), drawing partly on work of other authors in geography (e.g. Harvey, 1989) and social theorists from other disciplines such as Lupton (1993), Giddens (1994), Lash et al. (1996). Beck argues (1999: 56–79) that many of the processes generating important risks for modern society operate at the global scale. Examples of these 'world risks' include international terrorism, global circuits of capital, global warming, rapid diffusion of new infectious diseases. The conventional methods that societies had developed for managing risk, such as limitation and control of hazards, systems of insurance and compensation, advances in medical science, have meant that in other respects our lives are much more secure and comfortable than at earlier times in history. However, these conventional strategies are not able to prevent or mitigate the kinds of 'world risks' we now face and we are left with a sense of uncertainty and threat which generates a high degree of anxiety in modern society. Beck also argues that conventional views about the kinds of knowledge that can help us counteract risks in society are in question, especially in the face of world risks. He contrasts conventional ideas about 'linear knowledge' (summarised in Figure 1.4) with non-linear knowledge. We might have believed in the past that, to cope with risk, we could rely on linear knowledge, provided by a consensus view, developed among a limited circle of experts, which could provide a degree of certainty about risks and the likelihood of unintended consequences of human activity. Now, according to Beck, we are faced with a much more unsettling, non-linear view of the knowledge available to help us cope with world risks. This implies that far from a consensus view, there is a contested, unresolved field of knowledge debated among a much wider, ill defined set of 'experts' who do not all agree. Not only do we lack clear answers to questions about the risks we face, but we do not even know what these risks are and what sort of knowledge we need to address them.

linear unawareness in *not central* to reflexive modernisation	non-linear unawareness of various types is the *key problem* in transition to reflexive modernity
closed circles of experts acting on knowledge	an *open multiple field of competitors* acting on knowledge
knowledge is based on *consensus* of experts	dissent and conflict over rationality and principles between various networks of people: *contradictory certainties*
homogeneous expert groups discuss 'unintended consequences'	'we don't know what we don't know'

Figure 1.4   Linear and non-linear theories of knowledge, after Beck (1999)

Source: Curtis, 2008, p. 295.

Lupton, writing from a sociological perspective (e.g. Lupton, 1999; Lupton and Pedersen, 1996) and Bickerstaff (2004) Bickerstaff and Walker (2003) from a geographical point of view, are among many authors who argue that in light of these kinds of arguments we need to recognise that risk is socially constructed as much as objectively and scientifically determined. Also our understanding of risk is closely bound up with social relations of power and resistance in societies. For example, many of the criteria we use to make judgements about and carry out surveillance of 'safety' of levels of air and soil pollution, genetically modified crops, exposure to bacteria in foodstuffs are variable from one part of the world to another. Even within one national jurisdiction, the criteria for judging risks will be contingent on a number of other factors. Thus for example, practical judgements about the importance of risk posed by chemical contaminants in soils depend on what use is made of the ground and by whom. In addition, risks of human activities often have to be weighed against benefits of that activity. For example, a community might tolerate (even tend to deny the existence) of an uncertain degree of risk associated with environmental pollution from and industrial activity which provides employment and economic security for many people in the town. Furthermore, as discussed in Chapter 4, aspects of the social environment may be valued above physical environmental attributes within local communities (Wakefield and McMullen, 2005).

Pedersen and Lupton (1996) have reviewed arguments that these perspectives on risk have profound implications from a public health perspective, for the way that we view the health related risks to which we are exposed, the ways we interpret advice from 'experts' about these risks, and the degree that we feel ourselves responsible for risks we take or able to control the level of risk towards which we are exposed. The psychology of risk perception and the psychological impacts of such socially constructed perceptions of risks in the social and physical environment, as well as the mental health effects of actual experience of hazards, are topics for further discussion in Chapters 3 and 5 of this book.

Efforts by society to manage risk also have implications for socio-geographical relations of power and discipline. For example O'Byrne and Holmes (2007) use frameworks developed by Lupton to illustrate, in the context of preventive work of nursing professions, how social construction of risk and individual responsibility for trying to control risks operate through processes of social control and regulation postulated by authors such as Michel Foucault (1977). They discuss (O'Byrne and Holmes, 2007: 99) how "Risk becomes...[a]...yardstick by which citizens can be ranked as good or bad. Individuals who engage in activities that may put them or others at risk are bad citizens who must be made aware of their status through health related mass media campaigns. Furthermore...individuals are persuaded to submit their bodies to the gaze of an expert who will perform a more individualised assessment...". In Chapter 7 there is discussion of the ways that social and clinical constructions of risks associated with mental illness have led to the creation of spaces of surveillance and control of mentally ill people.

Emotional geographies: development of ideas about 'spaces of affect'

Going beyond the issues just discussed concerning risk, anxiety and trauma, this book also engages with some of the research on 'emotional geographies' and examines the significance of this expanding field of the geographical literature for mental health and wellbeing. The literature on 'geographies of affect' has elaborated on the links between feelings, place and space, and also complemented geographical research on the body, by considering the emotional response to different bodily states. Several commentators on this theme (e.g. Thrift, 2002, Bennett, 2004; Thein, 2005) have remarked that geographers have in the past been rather tentative about engaging with emotions as a 'valid' topic for research, but that more recently the significance of emotion as a force in socio-geographical processes has been grasped with more confidence.

Thein (2005) provides a useful explanation for a geographical audience of the idea of 'affect', drawing on ideas from Freud (1991) (first published in 1915), Matthis (2000) and Sedgewick (2003). The concept spans ideas about fundamental human impulses or 'drives' which influence our emotions (often subconsciously) and more organised and conscious thoughts and feelings about our interactions with other people and things in the world around us, as well as with our instrumental goals and aspirations. The geography of 'affect' is thus concerned to express the motives and expression of our feelings and emotions as they are manifested and constituted in place and space.

Geographers including Thein (2005) and Bennett (2004) argue for a view of emotion and affect as socially constructed, connecting to ideas developed in sociology (e.g. Lupton, 1998). This perspective emphasises the importance of the social environment for individual emotional experiences. Bennett (2004: 414) discusses the idea of emotions as "created and understood in the context of relationships with others and meaningful through discourses, language and signifiers." She cites research in geography, sociology and anthropology exploring the ways that social 'rules' prevailing in particular settings govern the emotions we expect (and are expected to) feel and express in different settings.

Geographers such as Anderson and Smith (2001) have argued that emotional geographies have considerable relevance for political and social processes, so that they have real significance for social policy as well as intrinsic 'academic' and personal interest. Thrift (2004) elaborates on this theme. His argument about the need to incorporate ideas of 'affect' into geographical research on urban society also offers some valuable reflections from the point of view of this book. Thrift draws on ideas from writers in various fields of philosophy and science, such as Spinoza, Deleuze, Darwin and others, to identify different ideas of 'affect'. These include the idea that affect might be 'a set of embodied practices' – positing an indivisible link between feeling and behaviour, in which context is crucial because 'the source of emotions come from somewhere outside the body itself' (Thrift, 2004 p. 60). A second interpretation of affect could be in terms of a sort of 'drive' or power of action (joyful, positive emotions enhance power of action, while sorrowful, negative emotions have the reverse effect). Another view of affect could be seen as a collective parallel to ideas about 'emergent' aspects of individual 'states of mind' from neuroscience, suggesting that affect 'emerges' from the combination of myriad interactions between individuals. Thrift has also referred to Darwin's theories about evolution and explored the idea that while some aspects of affect may be 'culture bound' others may be rather universal to humans and may have been developed because of their value in terms of evolutionary advantage for the human species. Thrift has argued that an emphasis on these aspects of affect would point to the development of certain avenues in geographical research: a stronger emphasis on affect as a part of socio-political processes in urban society, for example, through discourses relating to self expression and choice; greater attention to the mediatisation of politics which relies significantly on affect; research on aspects of performance through which political ideas are conveyed in order to influence, in ways that are not restricted to speech and the written word; and the design of urban space designed to provoke particular emotional responses allied to socio-political objectives. These ideas are considered in relation to mental health geographies especially in Chapter 6 of this book.

There also is a growing body of geographical research, invoking ideas about geographies of affect and emotion, which focuses more specifically on mental health and health care and on therapies (for example, reviewed by Bondi (2005) and this research is discussed later in this book (especially in Chapter 7).

Theoretical challenges for research on 'states of mind': *mind–body dualism and embodiment*

In daily life we often address questions about environmental factors relating to mental health and wellbeing in 'common sense' ways, as though mental health can be readily distinguished from physical health. Much of the research on 'mental illness' also proceeds without any explicit acknowledgement that mental illness, or psychiatric disorders are indistinct and problematic categories. This conceals some very complex issues about what we understand by 'states of mind' or by 'mental' as opposed to 'physical' illness. These are matters that have been debated in various academic disciplines for centuries, and it is certainly not the aim of this book to resolve them. Rather, this discussion aims to show how a geographical perspective fits into these debates, and how it may be useful as part of an interdisciplinary effort to explore and better understand the complexity that they represent.

There is a long history of philosophical discussion, especially in western cultures, over whether the human body and the mind are distinct and whether such a distinction makes sense from a modern, scientific perspective. This debate has significance in many fields of knowledge and it is outside the scope of this book to offer a comprehensive review. Furthermore, it is pertinent to note that some authors have argued that the whole issue is a sort of philosophical 'red herring'; a digression that is unhelpful for those wishing to understand human experience and our relationship to the world around us (Read (2008).

However, it is useful here to consider how far it is reasonable to concentrate on mental health specifically, as opposed to human health in a more general sense. The discussion here considers particularly how the debate has been taken up by authors working in psychiatry and psychological medicine, and in branches of geography that are concerned with human health and wellbeing. Different interpretations draw on two divergent points of view which are often presented in terms of the Cartesian, 'dualist' view of body and mind, as opposed to the phenomenological perspective.

The Cartesian perspective derives more or less directly from the writings of the 17th century philosopher René Descartes (1596–1650). His work has been published for example, in English translation as: Descartes (1985), and as, Descartes (2002) in a useful collection of readings on philosophy of mind edited by Chalmers (2002). Descartes' seminal discourse seeks to justify making a logical distinction between mind and body which would enable a rational view that the human body and brain are distinct from the human mind or 'spirit'. For example: he argues as follows (in this English translation of his 'Sixth Meditation meditation on *The Existence of Material Things, and the Real Distinction between Mind and Body* published as Descartes, 2002: 18–19):

> The first observation I make...is that there is a great difference between the mind and the body, inasmuch as the body is by its very nature always divisible, while the mind is utterly indivisible. For when I consider the mind, or myself in so far as I am a thinking thing, I am unable to distinguish any parts within myself.... I understand myself to be quite single and complete. Although the whole mind seems to be united to the whole body, I recognise that if a foot or arm or any other part of the body is cut off, nothing has thereby been taken away from the mind.

The analogy that is often cited is of the body as 'machine', functioning in a rather mechanical way, and the mind as the 'ghost in the machine', which wills bodily behaviours and makes one conscious. This leads to the idea of 'mind-body dualism' and the view that there must exist some 'substance' comprising the human mind, which is distinguishable from the human body, though it may be impossible to grasp its nature through scientific logic and methods. Benedictus de Spinoza (1632–1677), in contrast, was an early exponent of the 'monist' view that:

> ...(1) for each simple body there exists a simple idea that corresponds to it and from which it is not really distinct and (2) for each composite body there exists a composite idea that corresponds to it and from which it is not really distinct... (Dutton, 2006, section 4a).

This tension between dualist and monist perspectives was later pursued, for example in phenomenology (first articulated by philosophers writing in the late 19th and early 20th century such as Edmund Husserl, Alfred Schutz, Ludwig Wittgenstein, and Maurice Merleau-Ponty) (see, for example, a review by Read, 2008). This conceptual framework tends towards the view that the distinction between our perception of *phenomena* (denoting the outward appearance of things in our environment) and their existence is impossible to establish. This makes it very difficult to think about a separation between mind and body, and it also inspires doubts about whether there is a 'real' separation between the world around us and our perception of it. An example of these arguments which has been influential in geography, is the work of Merleau-Ponty (e.g. 2001 translation of the 1945 original), who explores the idea of human consciousness using a range of examples illustrating human experience of one's body and the world around it.

These include: the 'pre-conscious' reflex responses to physical stimulae enacted by the body without conscious thought; the perception reported by amputees of a 'phantom limb' which can be felt though it does not physically exist; the ways that we perceive space and movement; the ways that we learn movements or behaviour (for example the movements of a dance) that become 'habitual' so that, once learnt, we can perform them without thinking, and which may have a cultural or social, as much as a biological significance; and our capacity for empathetic awareness of other people. He writes about the intimate connection between our consciousness and our experience of the physical and social environment using the phrase 'being-in-the-world', for example:

> ...there appears round our personal existence a margin of almost impersonal existence, which can be practically taken for granted, and which I rely on to keep me alive; round the human world which each of us has made for himself is a world in general terms to which one must first of all belong in order to be able to enclose oneself in the particular context of a love or an ambition. ...the fusion of soul and body...the sublimation of biological into personal existence, and of the natural into the cultural world is made both possible and precarious by the temporal structure of our experience...Thus to sum up, the ambiguity of being-in-the-world is translated by that of the body, and this is understood through that of time. (Merleau-Ponty, 2001: 84–85)

He therefore argues for a blurring of the distinction between 'concrete' and 'abstract' accounts of human behaviour and mental states (Merleau-Ponty, 2001), as well as a perspective which understands consciousness in terms of one's experience of, and interaction with environment over time. This emphasis on the interactions and intricate interrelationships of human perception and environment, which make them conceptually inseparable, is relevant to much of the discussion in later chapters of this book.

Mind-body dualism viewed from the interface of psychiatry and philosophy

The tensions between these points of view are widely debated in terms of their relevance for psychiatry (and also in geography as discussed further on in this chapter). The significance of the 'Cartesian vs. Phenomenological' debate for psychiatry has recently been reviewed for, example, in several journals targeting researchers and professionals from psychiatric disciplines (e.g. Kendler, 2001, 2005; Schimmel, 2001a, 2001b; a collection of papers edited by Owen and Hardland, 2001; Bennett, 2007).

Several authors comment critically on a perspective from neuroscience which might suggest that it is only useful to focus on the physical and chemical malfunctioning (physiopathology) of the brain in order to understand psychiatric orders and how to treat them effectively. Thus, for example, Owen and Harland (2007: 106) introducing a special issue of *Schizophrenia Bulletin* on 'phenomenology and psychiatry in the 21st century' comment that it might be possible to dismiss phenomenology as of little value for modern psychiatry because the key to successful psychiatric medicine will be to focus on objective scientific knowledge about how illness arises from physiopathology of the brain. According to this view, subjective information (about 'states of mind' and human 'experience' of mental illnesses, seen as the outcome of 'faulty cognitive processes') might be argued to be unlikely to advance knowledge in contemporary psychiatry. Figure 1.5 illustrates an example of images of activity in the human brain when it is thinking about particular problems in laboratory conditions. This type of approach aims to examine human thought processes as physical and chemical activity in the brain and is the type of research that might also help to identify faulty brain function associated with psychiatric illness. Similarly, Schimmel (2001a: 485–6) concedes that the human mind is "contingent on the functioning human brain" and "that mental events and brain events constitute a fundamental unity".

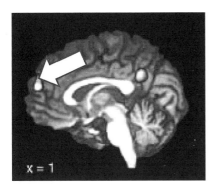 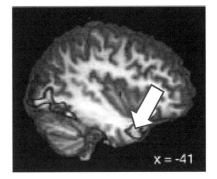

Figure 1.5   An example of brain scan images of activity in different parts of the human brain when stimulated by complex problems involving spatial and interpersonal relationships

Note: The arrows show parts of the brain which were most active when people were processing relatively complicated information about the presence and position of people. For example, when asked to think about the following question in relation to a given scenario: 'Nina sits in front of Peter who sits right of Thomas. Is that true?'

Source: Abraham et al. (2008), p. 446. Reprinted from *The Lancet* with permission of Elsevier.

However, these authors also argue that while such biomedical knowledge is certainly important, psychiatry also requires some knowledge and understanding of the human experience and 'state of mind' in order to form a comprehensive and meaningful picture of the nature and the causes of mental illness. Owen and Harland (2007, p. 106), point out that neuroscience has not been able to offer complete explanation or effective treatment for all mental disorders and, furthermore, psychiatric disorders can only be identified through knowledge of subjective experience of suffers, so that "Ignoring the 'what is it like' of a mental illness – its phenomenology – risks undermining the very objectivity of psychiatry". Andreasen (2007) reviews the history of the *Diagnostic and Statistical Manual of Mental Disorders (DSM-III)* emphasising the importance for the practice of psychiatry of information about signs and symptoms of mental disorder that were not based on biomedical measurements but on patient's 'internal subjective experiences' of their symptoms recorded by psychiatrists.

Schimmel (2001a: 485–6) also argues that current knowledge from neuroscience does not resolve the question of how we can understand the human mind and individual perception because "Any attempt to explain the subjective essence of a mental state or process in terms of a brain state or process must inevitably fail as it leaves out the mind that we seek to explain". Thus, "[e]ven if we wish to avoid a dualist view we seem to be stuck phenomenologically with a sense of duality; we simply experience mind and body as distinct". Schimmel (2001b) elaborates this point in a discussion of the apparent 'causes' of different psychiatric disorders (conditions which are medically recognised, can be diagnosed and may be medically treated, or at least may be alleviated to some degree by medicine and therapy). Some conditions, he argues, can be seen as arising from traumatic damage to the brain, due to head injury, or deterioration of the functionality of the brain (as in Alzheimer's disease). Neuroscience of the physical brain is a useful way to understand these causes of mental illness. However, he argues that the fundamental causes of a disease like schizophrenia are less well understood in purely biomedical terms; the symptoms may be caused by neural malfunction in the physical brain, but this can apparently be triggered by factors including social stressors which are initially experienced 'as mental events' which lead to physical changes in the body. Similarly a common cause of depression is bereavement, which is first experienced mentally as a loss of a social relationship. Schimmel argues that it is difficult to understand the cause of such depression and treat the patient effectively without acknowledging that the patient's 'state of mind', influenced by a social loss, is an important element of the disorder, even though the disorder may be physically manifested from a neuroscience perspective as a chemical imbalance in the brain.

Authors such as Kendler (2001 and 2005), Basar Karakis (2006), Basar (2006) and Bullock (2006) explore a possible resolution of these arguments through a more complex conceptualisation of the neurological functioning of the brain. Their arguments echo, from a neuroscience perspective, theories about complexity and emergence discussed above. For example, Arshavsky (2006:

190) advocates: "...the hypothesis that the performance of cognitive functions, is based on complex cooperative activity of 'complex' neurons that are carriers of 'elementary cognition'. The uniqueness of human cognitive functions, which has a genetic basis, is determined by the specificity of genes expressed by these 'complex' neurons." In similar vein, Basar (2006) discusses the idea of 'whole brain work', which tries to envisage a process by which many single bio-chemical 'oscillations', simultaneously taking place in the whole brain may work together to result in human thought and perception.

Psychosocial factors and physical health

The discussion is further complicated if we introduce research which suggests that states of mind can impinge on the experience and causes of physical health. For example, research on the psychology of pain, including aspects such as placebo effects that seem to allow people to experience alleviation of pain as a result of taking 'medications' which they believe to be effective, but which for which there is no biomedical basis for a therapeutic effect. This research is considered in reviews and collected papers by Gordon (1996); Harrington (1997); Moerman (2002). These studies often seem to illustrate that our physical experiences are partly influenced by individual perception and experience of pain in response to a given stimulus is quite variable from one person to another. Furthermore, psychological distress may be manifested or expressed by patients in terms of physical symptoms, and the use of somatic imagery to express mental distress may vary from one culture to another. (Mallinson and Popay, 2007).

Furthermore, there is growing evidence and debate about the extent to which psychosocial variables impact on physical health. Research on 'psychosocial' risks for health involves conceptualising and testing empirically the links between psychosocial environment, mental wellbeing and physical as well as mental health outcomes. A statement of the theory from an epidemiological perspective is provided by two well-known proponents of this perspective Marmot and Siegrist (Marmot and Siegrist, 2004; Siegrist and Marmot, 2004). They define the 'psychosocial environment' as "the sociostructural range of opportunities that is available to an individual person to meet his or her needs of well being, productivity and positive self-experience" (Siegrist and Marmot, 2004: 1465). The argument runs that these aspects of the psychosocial environment influence an individual's psychological state (for example one's sense of self-esteem and self-efficacy) and that one's psychological state is related to physiological processes in the body, such as hormone levels. If sustained over the long term, these physiological changes can be associated with varying risks of physical health outcomes such as cardiovascular diseases. As discussed later in this book, there is a large body of evidence, some of it rather equivocal, concerning the relative importance of these 'psychosocial' pathways for health outcomes.

Of particular interest in this book is the way that the psychosocial perspective on health conceptualises links between the social environment and the person's individual psychological and physiological state. Frohlich (2007) discusses the literature on psycho-social correlates of health and points out that it implies that *social processes* at the level of groups in society are important for health related behaviour and psychological health at the individual level. This creates a tension in public health research in this field because of the need to simultaneously conceptualise individual psychological states and motivations and also social processes at the collective level of social groups. Because disciplines outside medicine such as social geography and sociology are closely concerned with the interaction between the individual and their social environment, they provide useful frameworks to tackle this agenda. The social model of public health provides one of the conceptual frameworks that help us to organise research on the psychosocial environment as it relates to public health, as considered below.

More generally, in light of the discussion above concerning the connections between the body and the mind, these theories about psychosocial factors and health present another set of arguments supporting the idea that it is unjustified to consider mental and physical and mental health as separate aspects of health, and they should more properly be considered as closely linked.

Geographical ideas about embodiment, the individual and their environment

Within geography, a parallel debate has re-emerged particularly strongly in recent years turning around ideas about the body and embodiment. These also problematise the notion that the physical body (and physical health) can be properly understood 'objectively' in a way that is distinct from social and psychological perception of the body and what constitutes a healthy body. Reviews of theoretical and empirical research on geographies of the body and their relevance to geographies of health have been published by Butler and Parr (1999), Longhurst (2001) and Parr (2002a; b). Geographical theorisation of *embodiment* frequently explores social constructions and psychological perceptions of the body and its environment in terms of 'spaces' with socially and individually determined meanings and significance.

One stream of geographical discussion about the body considers the relationship between the body and its wider environment, recalling some of the phenomenological debates considered earlier in this chapter. Ingold (2006) discusses the way that western cultures tend to view the human body as distinct from and surrounded by its environment. Other, more 'animistic' cultures view the individual more as integral to the 'mesh' of their environment. They also stress the ways that individuals move through their environments, rather than focusing on their static position within particular settings. (See also an earlier discussion by Seamon; 1979).

Some authors (e.g. Cresswell, 1997; Longhurst, 2001; Hall, 2000; Little and Leyshon, 2003) also discuss the materiality of the body, with an emphasis particularly on the extent to which our views of the body are socially constructed (and therefore variable from one social setting to another). Longhurst (2001, p. 5–6) considers the following arguments: that bodies are 'real' and at the same time are socially constructed and are "...always in a state of becoming...The question 'what is a body?' can only be answered by 'locating' bodies...". She explores whether focusing on the body changes the way that we think about social relationships in space. Longhurst explores these ideas through a series of examples of human bodies in particular conditions, including pregnant women in public spaces, men's bodies in private spaces such as toilets and bathrooms, male and female bodies in the workplace. Through discussion of these examples, Longhurst develops a discussion about 'body-space relations'. The examples are used to show that "bodies and spaces are neither clearly separable nor stable" (Longhurst, 2001: 7). Similar arguments may be applied to mental states as they relate to geographical settings (e.g. as discussed in Chapters 6 and 7).

An important part of the geographical agenda on embodiment concerns the human body interpreted in terms of social power relationships (for example, the collection of papers edited by Nast and Pile, 1998), and a significant contribution to the geographical literature on the body has been made by feminist researchers investigating gendered social relationships. Longhurst (2001) also reviews phenomenological, psychoanalytic, and 'inscriptive' approaches to understanding the body as well as geographical discussion about the Cartesian dualist perspective. She identifies streams of research in geography that have deployed ideas from psychoanalytic theory, and social theories of power, consumption and commodification, showing how these have helped geographers understand how bodies are constructed as masculine or feminine, weak or powerful, as expressions of social differences and as manifestations of our needs, desires and emotions. Cresswell (1999), considers the historic case of female 'tramps' and what we can learn, by studying how public discourse of the time represented these women's bodies, about the relationships between spatial mobility of the body, social control of space, notions of propriety and 'appropriate' roles and settings for female bodies. Of significance for the content of later chapters in this book, concerning social and geographical marginalisation of people with mental illness, is Cresswell's discussion of the unease and sense of insecurity created in American society in the late 1800s and early 1900s by the presence of tramps generally, and especially by the unconventional and anarchic appearance and behaviour of female tramps. These studies emphasise the interrelationships between emotional, psychological and physical dimensions of social relationships which are important for mental health (as discussed in Chapters 4, 5 and 6).

There is emphasis in this field of geography on the role of the body in *performing* socially significant roles and in self expression, which is considered in more detail later in this book. Little and Leyshon (2003) present a discussion of relations between body and rural space, using examples from the UK. They consider 'performance' as an idea to interpret the importance of the body for every day practices which include expression of personal and social identity and social position and for carrying out economic processes of production.

There is also a significant emphasis on the human body in health and medical geography. Attention is particularly given to the way that medicine has established its authority in respect of human health by establishing a particular, biomedical view of bodily and mental states, described as the 'medical gaze' by Michel Foucault (1967, 1973). For example, Philo (2000a) describes Foucault's book *The Birth of the Clinic* as an "unknown work in medical geography, because of Foucault's use of the idea of corporeal *space* in his account of the 'archaeology' of medicine". Reviews by Parr (2002a, 2002b) discuss how this dominance of medical professionals in the deployment of this 'medical gaze' is perhaps being eroded by growing availability of information on the internet that facilitates self diagnoses by people without medical training. The psychological implications of the power relationships between doctors and their patients are particularly relevant for the discussion in this book. Indeed, more generally, the ways that ill bodies (and sick minds) are perceived and treated in different settings such as hospitals, clinics, public spaces and the home are also frequent subjects for research in medical and health geography, taken up particularly in Chapter 7. Geographies of health also engage with discussion about the 'social construction' of healthy and unhealthy bodies and minds in wider society. Issues of stigma attached to particular bodily or mental states which are treated as 'deviant' and 'dangerous' are important for the geography of mental health.

Broadly, therefore, we see that geographical perspectives view the body as a 'corporeal space' which is 'discursively produced' and which expresses social power relationships operating through social institutions which give the body its social meaning. Human geographers often emphasise also that the social significance of bodies is often contingent on the social construction of the wider setting in which they are 'located'. This work also implies a strong link between bodily state, emotion and mental wellbeing, which is particularly important for this book.

The social model of public health

Two threads of the arguments emerging repeatedly in the theoretical literature discussed so far are that (a) human mental and physical health comprises aspects that are socially constructed as well as biologically determined and (b) that there are close links between one's individual characteristics and one's social and physical environment, which have significance for mental and physical health. These are exemplified in detail later in this book, but here it is appropriate to mention that from a public health perspective the *social model of health* provides a framework which focuses particularly on these connections between environmental factors, individual factors and human health. This conceptual model is frequently represented in the public health literature by an 'onion' diagram like Figure 1.6. This model, proposed by Whitehead and Dalgren (e.g. Whitehead, 1995), places

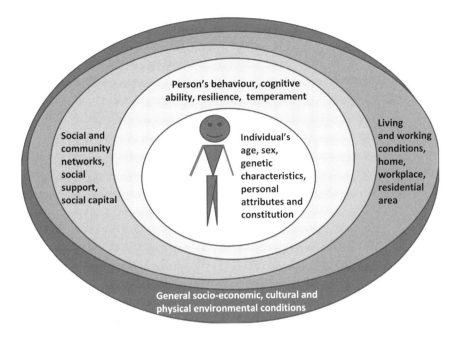

Figure 1.6   A 'Social Model of Mental Health' based on the 'social model of health' proposed by Whitehead and Dalgren (Whitehead, 1995)

individuals and their individual characteristics within their wider environment, made up of both social and physical aspects. The diagram is meant to summarise the 'wider determinants' of health associated with attributes of the social and material environment, and not restricted to medical care. This perspective is consistent in many ways with the approach taken in this book.

Varying definitions and measures of 'mental health' and wellbeing

In this book, which takes a primarily geographical perspective, discussed below, we are often concerned with measures of public mental health, suited for assessing and comparing populations (as opposed to making clinical diagnoses of mental illness in individual patients). (See Box 1.3 for a summary of some key terms used for techniques of measuring population health). Since the book takes a socio-geographical view on mental health and is concerned with good health as well as illness, aspects such as wellbeing and quality of life will be considered, as well as psychiatric definitions of medically recognised disorders. Various measures of mental health and illness will be referred to, which reflect these different dimensions

**Box 1.3   Measuring ill health in populations: plain language definitions
               of some key terms**

Text in italics is sourced from the University of Cambridge Dictionary of
Epidemiology, http://www.geocities.com/HotSprings/3468/gloss96.html
which also provides plain language definitions of other epidemiological
terms.

These notes are intended for readers who are not specialists in epidemiology.
More precise and detailed definitions and explanations the calculation to
produce these indicators are provided, for example, by Silman (1995);
Lilienfeld and Lilienfeld (1980).

Health of populations is often measured in terms of **mortality** (*death*)
or **morbidity** (*a state of ill health produced by a disease*). Mortality and
morbidity may classified in 'cause specific' categories (attributed to specific
diseases that are medically diagnosed). The 'level' of disease in a given
population may be measured by:

Prevalence – the proportion of a population that has a disease at a particular
time;

Incidence – the number of new cases of a disease that occur in a population
over a given time period.

Burden of disease in a geographically defined population can be measured
in terms of:

Counts (absolute numbers) in a population of deaths or people that have
an illness. This indicator may be useful for some purposes, but is usually
not suited for comparison of populations in different places where the total
population sizes are dissimilar;

Crude Rates of mortality or morbidity express numbers of deaths or
disease cases as a proportion, relative to the total population size. They
can be used for rough comparison of ill-health in different populations but
must be used with caution as they do not control for the age and sex profile
of different populations.

Age /Sex standardised rates of mortality or morbidity are comparative indicators that may be calculated using direct standardisation techniques. These 'project' the death/disease rates prevailing in each age/sex group of a population of interest onto the age and sex profile of a standard 'reference' population. If age/sex standardised rates of mortality or morbidity are calculated for populations of different areas using the same reference population, the effects of differences in population size and area age/sex profiles are screened out of these measures. This makes them more useful than crude rates as indicators for comparing underlying differences in illness or death for populations in different areas.

Age/sex standardised ratios of mortality or morbidity are comparative indicators that may be calculated using indirect standardisation techniques, which compare the observed count of deaths or diseases in an area with the number that would be 'expected' if the area had the same rates of events of death/disease in each age/sex group as those prevailing in a standard 'reference' population. The indicator is often expressed as a percentage ratio of observed to 'expected' events, so that 100 represents a level of health similar to the reference population, while a value different from 100 shows that health in the area being considered differs from the reference population. For example a standardised mortality ratio of 90 shows mortality to be 10% lower (better) than the reference population while a value of 110 denotes mortality that is 10% higher (worse) than the reference population. When they are calculated for a number of areas using the same reference population, Standardised Mortality/Morbidity ratios are useful for comparing illness or death in geographically defined populations because they control for area variation in the age and sex profiles of the populations.

of mental health status, across the spectrum from severe, psychiatrically defined illnesses through minor psychiatric illness and measures of 'distress' to more positive aspects of good mental health and wellbeing.

Measuring mental ill health

For many people, the term *'mental health'* first brings to mind ideas which are really more accurately notions about *mental illness* (being concerned with disturbed or troubled mental states which are considered 'abnormal' and may in some way limit routine functions of daily life). Indeed, much of the debate about 'mental health', in popular media and in discourses of fields related to psychiatric medicine, takes an even narrower view, referring to the presence or absence of *psychiatric disorders* that have a medical diagnosis (Figure 1.7). In fact, it is often the case that only

organic mental disorders (e.g. Alzheimer's)
mental disorders due to psychoactive substance use
schizophrenia/schizotypal and delusional disorders
mood (affective) disorders (includes bi-polar affective disorder)
neurotic, stress-related somatoform disorders (anxiety, phobias,
obsessive-compulsive disorders)
behavioural syndromes (e.g. eating, sleeping disorders)
adult personality disorders
mental retardation
disorders of psychological development (e.g. speech language)
behavioural/emotional disorders in young people (e.g. hyperkinetic)

**Figure 1.7   Psychiatric disorders (as defined in International Classification
of Diseases Version 10 (ICD10): Chapter V Mental and
Behavioural Disorder (WHO, 2007)**

a limited range of psychiatric disorders are given attention such as: psychotic disorders including schizophrenia (especially the relatively rare cases which give rise to violent behaviour on the part of the sufferer); 'severe' cases of bipolar disorder or depression (associated, for example with high risks of suicidal behaviour) or (less commonly discussed) severe loss of mental function in advanced cases of Alzheimer's disease. Perhaps these rather extreme aspects of mental disorder attract most attention because the individuals suffering from them require particularly large investments of resources for medical and social care in the community, as well as institutional provision for the most serious cases. Also, some behaviours that may be associated with these conditions are clearly different from what is socially defined as 'appropriate' or 'normal', and therefore tend to attract more attention, condemnation and stigma. In fact, a wider range of mental illnesses need to be considered, including: less severe forms of depression; anxiety and stress related conditions; phobias and obsessive-compulsive disorders and behavioural syndromes such as eating and sleeping disorders. These less severe conditions make up a large part of the burden of mental illness in human populations.

From perspectives in health geography or sociology of health one might take a more 'relative' view of such diagnostic criteria, considering them not as fixed, undisputed definitions, but more as the constructions of a particular, influential social group (certain psychiatrists in western countries). At least historically, there has even been disagreement amongst psychiatrists; Andreasen (2007), for example, considers the controversies over the formulation of the Diagnostic and Statistical Manual of Mental Disorders version III (DSMIII) classification which preceded the fourth and current version (DSMIV), indicating differences of opinion even within the medical profession about how to recognise psychiatric disorders. Yevelson et

Box 1.4   Measuring symptoms of depression in population surveys

Instruments measuring 'probability' of depression include the CES-D and the GHQ.

These scales do not produce a clinical diagnosis of depression for an individual, but they have been validated against diagnosed cases of depression so it is possible to estimate the probability that an individual may be clinically depressed. Used in community based surveys they provide a general view of the likely prevalence of depression in the population. Some of these measures require clinically trained personnel to administer them, while others do not require special training, so that they are suitable for general health surveys of the population. It is usually important that the respondent answers all the question items on the scale, so that responses to selected items like those below cannot be used in isolation.

The Centre for Epidemiologic Studies Depression Scale (CES-D) asks respondents about their experience of various symptoms of depression; examples of items are as follows:

You felt sad.
You felt like you could not get going.
You felt lonely.
You felt bothered by things which don't usually bother you.

The *General Health Questionnaire* is available in various versions; the 12 item questionnaire is widely used in community surveys. People area asked to answer 12 questionnaire items about 'how their health has been in general, over the past few weeks' and to indicate whether these items more or less than usual. Illustrations of the items include;

Have you recently:
Felt constantly under strain?
Been feeling unhappy and depressed?
Been feeling reasonably happy, all things considered?

al. (1997) discuss the differences between the *National Traditional Classification*, of psychiatric disorders being used in Soviet countries in the 1990s, and the International Classification of version 9 Diseases (ICD9) classification widely used in western countries at that time. These examples highlight the potential for differences in expert opinion about what constitutes mental illness, which persisted until relatively recently, when clinicians generally adopted the DSMIV and ICD version 10 classifications. It is sometimes argued that the dominant medical view is associated with western values concerning mental health and wellbeing. Later chapters of this book consider cross cultural differences in socially constructed ideas of mental health and wellbeing.

Psychiatric conditions generally are estimated to present a significant burden of ill health internationally. For example, World Health Organization (WHO, 2001) estimated that globally 31 per cent of years lived with a disability were attributable to mental and neurological disorders. The proportion varies from over 40 per cent in Europe and North America to around 18 per cent in Africa where many physical diseases are highly prevalent and cause relatively high rates of premature illness and death. These estimates are quite debatable because of the difficulties of measuring health variation across the large number of countries included, and the challenge of calculating the global burden of ill health over the whole lifecourse, especially as the estimates are not based on direct measurement in the population in all cases. Nevertheless they underline the significance of the burden of psychiatric disorders in countries around the world.

More precise studies in specific countries have estimated prevalence of serious conditions using more robust population surveys. For example, Jablensky et al. (2000) report a survey in Australia showing that the prevalence in a typical month of schizophrenic disorder for the population aged 18–64 was only of the order of 4–7 cases per 1,000 people. More widespread in the population are milder, or less 'conspicuous' forms (episodes of milder anxiety and depression or alcohol dependency, panic attacks) which may be transient rather than chronic but can have important impacts on people's daily lives). Alonso et al. (2004a) report levels of common mental disorders of this type in a sample of over 21,000 people in 6 European countries. They report that 9.6 per cent of the sample had experienced at least one episode of such mental disorders in the previous 12 months and that 25 per cent had experienced one or more episode in the course of their lives. Furthermore, over a twelve month period people with mental disorders reported rates of working days lost that were over twice the rate for people with no mental health or chronic physical problems, so that these conditions seem to have a significant association with disruption of everyday roles (Alonso, 2004b). WHO (2001: 24) reports national estimates of the prevalence of common mental disorders among primary care patients from various countries, ranging from 9.5 per cent in Nigeria to 52.5 per cent in Chile. Hu (2006) reports an international review of estimates of the economic costs to societies of mental illness which also demonstrate varying, but often quite significant economic burdens due to loss of economic activity and costs of care.

Measuring mental wellness, happiness and wellbeing

This book will also refer to literature relating to growing preoccupations in contemporary society with 'happiness' and wellbeing (e.g. Morris, 2004; de Botton, 2006). The idea of happiness is obviously a very personal and 'subjective' one (see

Figure 1.8 for some rather varying views of what counts as being 'happy'). Also, as we will see in the course of this book, happiness, good health and wellbeing are not the same, and they are not even necessarily linked. Carlisle and Hanlon (e.g. 2007) have conducted wide ranging reviews of different perspectives on wellbeing and they identify perspectives drawing on philosophical sources which distinguish between 'happiness' and well-being as 'part of a flourishing human life'. Fleuret and Atkinson (2007) have reviewed measures of wellbeing from a geographical perspective. They note that wellbeing is assessed in ways which variously focus on theories about needs and the extent to which these are met; ideas about relative wealth, standards of living and expectations for attainment in life; and concepts of capabilities. They point out that these ideas of wellbeing are very interesting for geographers for several reasons. First, these ideas are open to considerable variation in interpretation from one setting to another and among different societies. Second (as discussed above), environmental variables will play an important role in experience of wellbeing. Third, in different settings ideas and information about wellbeing are taken up variably in terms of policy, social action and individual endeavours.

As <u>materialistic</u>:
Happiness is...a large income
Attributed to Jane Austen , by Morris, 2004, p. 165

As an <u>attitude</u>:
Happiness is not the number of blessings, but only our attitude towards them
Attributed to Alexander Solzhenitsyn, by Morris, 2004, p. 152

As <u>sensual</u> :
*Happiness is
un-repented pleasure .*
**Attributed to Socrates
by Morris, 2004, p. 164**

As <u>rationality</u>:
*Happiness is...
not believing in
miracles*
**Attributed to Johann
Wolfgang Von Goethe
by Morris, 2004, p. 167**

As <u>tranquillity</u>:
Happiness is...neither wealth nor splendour but tranquillity and occupation
Attributed to Thomas Jefferson by Morris, 2004, p. 169

Figure 1.8   Expressions of 'the nature of happiness'

Source: Morris (2004) *The Nature of Happiness* (London: Little Books Ltd).

Box 1.5   Illustrative items from the WHOQOL questionnaire

Below are sample items from the WHOQOL questionnaire which has 100
questions in total. The instrument covers 25 different 'facets' of quality
of life, each of which includes several questionnaire items. Those shown
here are illustrations only, and the 'facet' to which they relate is shown in
parentheses.

How much do you enjoy life? (positive feelings)
Do you have enough energy for everyday life? ('energy')
Do you feel happy about your relationships with your family members?
('personal relationships')
How much are you able to relax and enjoy yourself? (participation in
recreation/leisure)
Do your personal beliefs give meaning to your life? (Spirituality/religion/
personal beliefs)
How much do you value yourself? (self esteem)

Source: WOHQOL Group (1998) Table 4, pp. 1576–7.

Internationally, many different measures of 'quality of life' are used, some of
which relate more particularly to physical or mental impairment, while others
include more positive aspects of wellbeing (e.g. reviewed by Bowling, 1997).
One scale which has been very widely used to assess psychological wellbeing is
the *SF36 Health Survey* (Ware and Sherbourne, 1992; Ware, 2009). This includes
items which relate to both physical and mental health. The 'mental component'
includes items relating to 'mental health', emotions, social functioning and
vitality. Another scale which has been specifically designed for use in a number of
different countries and cultures is the *WHOQOL* questionnaire disseminated by the
World Health Organization (e.g. WHOQOL Group, 1998). The full questionnaire
includes 100 items. Box 1.5 shows illustrations of these which have been selected
as being more orientated toward 'positive' aspects of health. Many of the other
items ask about difficulties and dissatisfaction with life. (there is also a short
version, *WHOQOL-BREF*). It addresses multiple dimensions of wellbeing and can
be used to produce information about 25 different 'facets' of wellbeing, grouped
into physical, psychological, independence, social, environmental and spiritual
domains as well as a subset of questions relating to 'general' quality of life. The
WHOQOL SRPB Group (2006) argue that one of its strengths, compared with
other measures, is the inclusion of items which relate to spiritual health, which they
show to be significantly associated with other dimensions of quality of life. This
measure was developed by an international team working in a variety of cultural

settings, which gives scope for cross-cultural comparison, while most of the other survey measures of psychological health and wellbeing were originally developed in western societies. For example Saxena et al. (2001) found that although the broad pattern of response to WHOQOL-Bref was similar in terms of the ways that different sets of questions grouped together into domains, there was some cultural variability in the importance attached to individual items on the WHOQOL-Bref questionnaire. This raises questions about cultural differences in understanding of wellbeing, which may contribute to international differences in wellbeing of populations as measured on this scale.

Conclusions: Agendas concerning spaces, risk and mental health

This chapter has shown that there is an ongoing debate, in various academic disciplines and in society generally, about how we understand mental health and wellbeing. There is also a philosophical debate over the nature of human psychological experience and whether mental and bodily states are distinguishable from one another. These arguments affect the ways that we view relationships between people and their environment.

This debate about mind-body dualism has been identified as an important issue for research on geographies of mental health. Several of the commentators considered above, arguing from a psychiatric perspective, critically examine the argument that the question of 'mind-body dualism' is no longer an issue, given advances in modern neuroscience. Some suggest that the debate is best advanced by the elaboration of more advanced theoretical frameworks concerning the complex physiological workings of the human brain that might more helpfully structure our thinking on the issue. There are also calls to extend research using social science perspectives such as 'relational' geographical theories about complexity, which are explored in this book. Some writers in the literature relating to psychiatry and geography also make a case for adopting elements of a phenomenological perspective to help us structure our thinking on the matter, and geographical literature using such concepts will also be considered in this book. The causal pathways that influence mental health and wellbeing are not identical to those that are important for physical health, even though they are closely linked Thus, while there are complex connections between 'mind and body', this chapter sets out a case for considering these connections through more specific study of processes relating to differences in mental and psychological aspects of health. Mental health at the individual and also the population level will be considered in order to identify key messages for policy and practice in mental public health, discussed at the end of the book.

The discussion in this chapter has set out the agenda for a geographical perspective on mental health which considers space and place as important for mental health variation and construes ideas about risks for mental health in terms of material, social and symbolic aspects of the human experience of the environment. To draw out these arguments the following chapters are organised to focus specifically on material, social and symbolic dimensions of environment as they relate to mental health and health care. Theories relevant to each dimension are explored separately in each chapter, but the structure of the book is not meant to imply that these dimensions are unrelated. Indeed, the discussion aims to show how they are linked. The book will review how our understanding of mental health is progressed by advances in 'relational' geographical approaches drawing on concepts of complexity, networks and space-time relationships in human geographies.

Notes for students

Key learning objectives

This chapter will help students in this field to gain knowledge of:

- the scope and motivation of health geography as it applies to mental health particularly;
- different theories concerning the nature of space and place that are relevant for geographies of health;
- connections between geographies of perception of space and place and ideas from psychology and neuroscience about environmental and spatial perception;
- approaches to measurement of mental health of populations and what these tell us about the burden of mental illness.

Introductory reading

For the material covered in this chapter key texts include the following (see also the guides to reading in Boxes 1.1 and 1.2).

For a 'classic' in health geography that still provides a sound basis for thinking about health geography, although it was published some time ago, and includes a chapter specifically on mental health, see: Jones, K. and Moon, G. (1987), *Health, Disease and Society: An Introduction to Medical Geography*. London: Routledge. For a more recent overview which also includes specific discussion of mental health, see: Curtis, S. (2004), *Health and Inequality: Geographical Perspectives*. London: Sage.

Overviews, with a large number of relevant references, that shows the development of geographical interest in mental health over time are: Wolch J. and Philo C. (2000), From distributions of deviance to definitions of difference; past and future mental health geographies. *Health and Place*, 6(4): 137–57; Parr, H. (2008), *Mental Health and Social Space*. Oxford: Blackwell.

Chapter 2
'Virtuous Landscapes': Therapeutic Material Settings

Summary

This chapter draws especially on ideas from geography and from environmental psychology to explain how our mental health is influenced by *material* aspects of our environment. The focus is therefore on physical features of the environment, and the ways that these are individually and socially perceived and interpreted in ways that are important for mental health and wellbeing. Key theories about the relationships between one's environment and psychological state are explored, especially theories of *biophilia*, *Attention Restoration Theory* and *topophilia*. Examples of empirical studies of the relationships proposed by these theories are discussed. These show that variability and change in the material environment are important for variations in mental health of individuals and populations. 'Therapeutic' settings, associated with relatively good health, or having potential to heal mental illness and distress are the main focus in this chapter. Therapeutic qualities of natural landscape elements are considered with special attention to trees and waterscapes. This chapter also explores aspects of built form which seem therapeutic. (Hazards in the material environment that are significant for mental health are discussed in Chapter 3.) The relationships between mental health and experience of the material dimensions of environment are seen to be closely associated with experiences in social and symbolic space.

Introduction: material environments and health viewed from a human geography perspective

This chapter taps into a core theme for geographers concerning the interactions between people and their physical environment. It explores the complex relationships between the material (physically tangible) attributes of our environment, the ways that these are perceived, experienced and interpreted, and how they relate to human mental health. We are concerned here with both 'natural' and 'fabricated' elements of environments and the emphasis in this chapter is on studies which consider these in specific settings.

Chapter 1 introduced some of the geographical perspectives that are important for the discussion in this chapter. There is a very long tradition in geographical research exploring the links between people and their material environment, which has been informed by a range of different theoretical perspectives. Theories with an older pedigree include ideas about 'environmental determinism'. The early phase of development of these theories in the period 1890–1920, is referred to by Judkins et al. (2008: 19) as the 'moment of environmental determinism'. At this time authors such as Semple (1911) were suggesting that human characteristics and the organisation of human society are fundamentally shaped by geographical and temporal variations in physical environments. More recent geographical research has questioned 'deterministic' ideas and placed more emphasis on the ways that physical conditions are manipulated and interpreted by humans in various ways, so that the links between physical environment and the 'human condition' are strongly contingent on social processes. However, it is also clear that human intervention cannot control and manage environmental processes completely. As discussed in Chapters 3 and 8 such ideas are being revisited and revised currently in response to emerging global environmental threats that are altering the way that environmental risks to human societies are understood.

A common theme running through all this debate is that there are interactions amongst the material and the social aspects of places, and the characteristics of individual inhabitants, which are important for many aspects of individual experience. Health geography places particular emphasis on the outcomes of these processes for human health.

Some of these processes seem to promote good health or health improvement. For example, Gesler (2003, pp. 1–6) proposes a 'therapeutic landscape' interpretation which defines a 'healing place' as:

> …a place that is conducive to physical, mental, spiritual, emotional and social healing…[and which is seen to] achieve a healing sense of place because several …related types of environments have been created there.

Gesler's conceptual framework of 'therapeutic landscapes' includes ideas about 'healing' and health promoting attributes of material landscapes, such as natural landscape features and design of 'natural' / built environments, and emphasises ways that their association with health depends on their *social* and *symbolic* importance, as well as their direct *physical* effects on the human body and mind. The therapeutic landscape framework shows how the presence of these therapeutic elements in particular places leads them to acquire a 'reputation' as 'healing places'. Experience of 'healing places' may help people to recover from illnesses including mental disorders, and may also help to maintain and promote good mental health. Understanding how and why these 'healing' and 'health promoting' processes operate is important practically, as well as theoretically, since, as discussed in Chapters 7 and 8, research on these issues helps to inform efforts to make environments 'healthier' for human populations generally.

Complementing this perspective on the benign impacts of certain attributes of landscapes, are perspectives which focus on the detrimental 'risks' to human health of other landscape attributes. These risks are partly associated with material elements that are 'absent', producing a lack of 'healing' elements in some places. Furthermore, some theories about 'landscapes of risk' for mental health concentrate on *present* material features which are harmful to our mental state, generating stressors which undermine and damage mental health and may cause, or at least exacerbate, mental illness. These are features which societies may seek to prevent, manage or at least mitigate in order to protect mental health of their populations. Discussion in Chapter 3 concentrates on these material hazards for mental health.

Material elements of therapeutic landscapes and landscapes of risk are unevenly distributed in space and they vary over time, so that they contribute to inequalities and trends in mental health and wellbeing. This means that a discussion of these landscapes has significance for social policies concerned with environmental justice and health inequality for different social groups. The discussion returns to these issues in Chapter 8.

The following sections of this chapter first review examples of research concerned with human responses to 'natural' aspects of places, in light of theories including biophilia, biophobia and restoration theory that help explain psychological responses to the 'natural' environment. The discussion then moves to a consideration of landscapes which are 'fabricated' by humans, and discusses our psychological responses to the design of built landscapes, reviewing research drawing on ideas such as topophilia, environmental psychology and landscape architecture and urban planning.

Natural landscape features and mental state

Many 'natural' elements in the landscape have significance for mental health and wellbeing. Few people in modern societies have frequent or prolonged experiences of entirely 'pristine' natural environments that are not in some way influenced by human activity, but most of the environments we occupy include elements that have not been built by humans and which constitute or recall what we think of as the 'natural world'. Natural features are dominant in 'green' and 'blue' spaces comprised of wild land-and water-scapes or open agricultural land and parkland, that are largely occupied by flora and by wild, and domesticated fauna, in a topography mainly comprised of exposed soil, rock and natural water systems. At the other extreme, given the increasing concentration of the world's population in urban areas, most people have more regular, day to day experience of 'incidental' natural elements within more generally 'built' environments. These include small urban gardens, or interior features of buildings including potted plants, water features, or domestic pets. Representations of natural environments such as pictures, or distant views though windows may be our most common forms of 'contact' with 'nature'. This increasing 'separation' from direct contact with predominantly natural landscapes, is a common feature of everyday experience for many human beings and is thought to have significance for our wellbeing. Furthermore, the following examples show that we have an ambivalent psychological relationship with these 'natural' features of landscapes. In some circumstances, we find them satisfying, reassuring and aesthetically pleasing, but the 'inappropriate' or 'damaging' 'intrusion' of the natural world into environments we create for ourselves can also be destructive, worrying (even frightening) and unpleasant.

Theories about 'benign' environments for human mental health

Various theories have been put forward to account for beneficial human reactions to material manifestations of 'nature'. Most of these were originally derived from traditions of thought rooted in 'western' cultures, and influenced by the dominant social groups within those societies. However, as shown below, research in these fields has begun to pay more attention to the socially contingent and diverse aspects of our relationships with 'nature'.

The theory of *Biophilia* (Fromm, 1973; Wilson, 1984) postulates "a fundamental, genetically based human need and propensity to affiliate with other life and lifelike processes" (Kahn, 1997: 1). The theory argues that our response to nature today is influenced by universal, inherited human characteristics, which would have conveyed primeval evolutionary advantages for the human species where it developed in Savannah landscapes. At the earliest stages of evolution humans would have been drawn to the elements that were necessary for survival such as fresh, clean water, edible plants and to settings where it was possible to catch and kill animals and fish for food, as well as shelter in places safe from attack by predators and other humans and protected from natural risks of severe weather, hot sun or extreme cold. According to the biophilia thesis, this would account for typical human preferences for 'natural' settings offering necessary resources for life and protection (especially waterscapes and 'open' green spaces, with shade and useful plants, as well as edible fauna). Human beings would benefit from a good understanding and knowledge of their environment; they are also thought to have become a successful species due to their relatively large brains, giving capacity to analyse and adapt to different environments, and giving them a sense of inquisitiveness that would have been advantageous, motivating them to explore and exploit new settings offering new opportunities. Thus natural environments that, on one hand seem understandable and accessible, but on the other hand offer potential for exploring new, unknown territory would be attractive. Biophilia therefore also emphasises 'legibility' and 'mystery' as complementary features of natural landscapes attractive to humans (e.g. Kaplan and Kaplan, 1989).

Similarly, Appleton's (1996) 'prospect and refuge' theory suggests that humans tend to prefer settings which offer protection, but also a commanding outlook on their surroundings. His discussion (Appleton, 1996: 52–72) analyses the imagery and symbolism of 'refuge' and argues that this is founded in a universal human need for places that offer security and shelter.

Figure 2.1   An example of a 'natural' environment interpreted through the theory of Biophilia

Note: Figure 2.1 shows an example of the kind of scene that might be expected to be attractive according to theories of biophilia. It shows a promontory of land extending into an open waterscape, with a path leading to the tip of the peninsula, just out of sight where an even wider view of the water might be had, we might feel drawn to walk to the very end of the track to find out. Although there are a few signs of human habitation in the distance, the scene is mainly comprised of 'natural' elements – trees, grass, open water which looks clear, unpolluted and likely to contain fish. The path is clear, flat and easy to walk along. In fact, the landscape elements are not all as 'natural' as they might appear. The picture was taken at the confluence of the Canal de Lachine in Montreal, Quebec, Canada, so the waterscape has been engineered to a degree. Also, the site is within walking distance of one of the largest cities in Canada.

The waterside park (Figure 2.1) at the junction of the St Lawrence river/Canal de Lachine Montreal, Quebec, provides access to a landscape which incorporates some of the 'natural' features which, according to theories of Biophilia and Attention Restoration, may be beneficial to psychological health. These include an open vista of water and sky, a negotiable route drawing one to a point further down the track which appeals to one's sense of curiosity.

Kahn (1997) reviews arguments by commentators on the biophilia thesis who have identified a number of criticisms of the theory. They argue that there are severe limitations to an interpretation which relies on such deterministic socio-biology, since people are not entirely driven by genetically encoded, primitive instincts, but also influenced by their own lifetime experiences and environment and by contemporary social influences of their social group. Also, as discussed in the following section, people have strong positive affiliations with 'un-natural' elements of modern day built environments which have little relationship to primeval landscapes and furthermore the natural settings that people find attractive do not necessarily have the appearance of Savannah landscapes. Furthermore there is evidence that many people have a 'biophobic' response so that they are repelled by 'natural' landscapes. Indeed, Koole and Van den Berg (2005) argue that in order to be attracted to natural landscapes, a person first needs to overcome a sense of fear of such places, since for many people there is a tendency to associate wilderness with death. In light of his review of these criticisms Kahn (1997: 21) discusses the case for 'mediated biophilia', such that humans have innate 'tendencies' toward biophilia but these are profoundly influenced by culture, experience, and the individuals' 'free will', which may reinforce or weaken these tendencies.

Another theoretical interpretation which may help us to understand the psychological response to natural landscape features is offered by Attention Restoration Theory (Kaplan, 1995). This theory argues that people respond positively to 'natural' landscapes because, when one is tired or stressed, these help restore a healthy state of mind and the ability to concentrate. Natural landscapes have 'soft fascination'; elements of interest that capture our attention without effort involved in their observation. Also contributing to their restorative effects are perceptions of natural landscapes as offering aesthetic beauty and tranquillity. Gibson's (1977) theory of affordances also connects to these ideas about the varying elements that different landscapes offer to people that are important for their psychological wellbeing.

Are 'green' and 'blue' landscapes good for our mental health?

Theories concerning the health benefits of natural landscapes have been tested in a large number of empirical studies, principally in the field of environmental psychology, but also in human geography. Some examples are considered here, selected to demonstrate the range of approaches used in this research.

Some studies aim to assess the associations between the 'stimulus' of natural landscapes and the psychological 'response' of interest by 'isolating' human research 'subjects' in laboratory conditions, designed to control for other factors that might affect their responses. Very often these sorts of study record responses to 'virtual representations' of natural environments (usually photographs or video clips). Attempts are made to measure human reactions to these stimuli in ways which minimise the 'subjective' aspects of self reported indicators of mental state. A typical example is a study by Ulrich et al. (1991), designed to test for the responses anticipated by Attention Restoration Theory. The participants took part in an experiment that involved watching a stressful film, which tended to produce a rise in blood pressure. Some participants were then shown further video film of natural scenes while others were asked to watch film of 'unnatural' scenes; motor

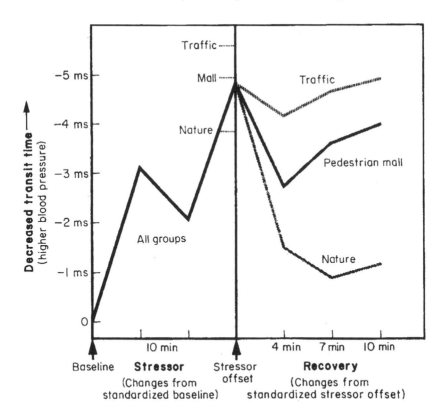

Figure 2.2   'Recovery' of normal blood pressure in relation to 'exposure' to film of natural and unnatural settings

Source: Ulrich et al. (1991). Reprinted from *The Lancet* with permission of Elsevier.

traffic or a pedestrian shopping mall. Those watching the natural scenes recovered normal blood pressure more quickly than the others, which was taken as evidence that the images of nature contributed most to 'restoration' of a more relaxed state. The study is also interesting because of the way that it demonstrates physiological responses to psychological stimuli, highlighting the strong associations between physical and mental health.

Other research on natural landscapes and human health uses an ecological strategy, testing associations between health outcomes and geographical information on proximity to green space. In a Dutch study, DeVries et al. (2003) related data on mental health from a national sample survey to geographical information for each respondent's residential area, measuring the 'green' and 'blue' space (areas occupied by vegetation and water) in proximity to their home. They found that those living in areas with more extensive greenspace, including gardens, woodland and agricultural land, had better levels of mental health measured on the General

Health Questionnaire. They found that this association was especially strong for people in less educated social groups, women working in the home and older people, and they suggest that this might be because, for these groups, access to green space is more likely to be restricted to the areas close to where they live. The authors of the study also comment that some of the associations they found might have reflected a general tendency for worse health in the most urbanised areas, but their findings nevertheless seem to support the argument that greenspace in the area fairly close to home and within walking distance (up to three kilometres in this study) may have significance for health.

A study by Mitchell and Popham (2008) examined the association between mortality data and average income for very small geographical areas in England in order to test whether this relationship varied according to the amount of greenspace in the area. In contrast to the Dutch study described above, Mitchell and Popham used an entirely ecological strategy, rather than data on individuals. Also they addressed a different question, as they were interested to see whether income *inequalities* in age and sex standardised mortality ratios among small areas were more pronounced in zones with less greenspace than in greener areas of the country. While they acknowledged that information on greenspace in their study might be acting as a proxy for other community variables, they argued that for some health outcomes, the disadvantages of poverty might be mitigated in part by access to greenspace for healthy physical activity. For all cause mortality and deaths due to cardiovascular diseases, there was an interaction such that in the greenest parts of the country, the mortality 'gap' between rich and poor areas was less than in regions with the least amount of greenspace. This would be consistent with the idea that greater levels of physical activity in greener areas helps to mitigate the impact of material poverty on health outcomes such as heart disease. However, this association did not hold for deaths due to self-harm, which relates more closely to mental, rather than physical illness. This aspect of their results may be partly explained by the excess of suicide mortality in rural, greener areas as well as inner cities (Middleton et al. 2008; and see Chapter 5 below) and the fact that rural health inequalities for less common causes of death are relatively difficult to assess using ecological data, especially in sparsely populated rural areas, because there is less differentiation in average wealth between areas than in cities and health statistics are based on small counts (e.g. Riva et al. 2009).

Other studies have inquired into informants' perceptions of their 'real' and direct experience of natural landscapes, and have used interview and questionnaire methods to record perceptions and attitudes to natural landscapes. Sugiyama et al. (2009) describe a community survey in Adelaide which explored links between green space and measures of self reported health. They collected information to construct a 'perceived greenness' scale based on responses to questions about access to parks or nature reserves, access to cycle or walking paths, presence of greenery or tree cover and canopy over footpaths and of 'pleasant natural features'. They investigated the association with mental health

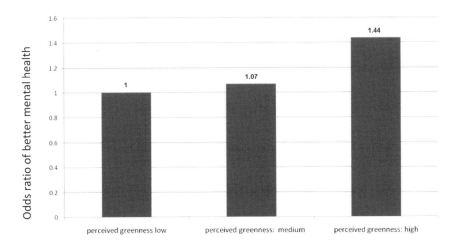

Figure 2.3   Community survey in Adelaide: odds ratio of having better mental health: people living in areas with medium or higher levels of 'perceived greenness' compared with those in areas where greenness levels were perceived to be low

Note: After Sugiyama et al. (2009) Table 4 (Estimates controlling for socio-demographic factors and whether respondents walked for recreation). Reprinted from *The Lancet* with permission from Elsevier.

measured on the 'mental health' components of the *Short Form Health Survey* questionnaire instrument. Figure 2.3 illustrates estimates of odds ratios for good mental health (reported by Sugiyama et al., 2009, Table 4), controlling for socio-demographic factors and whether respondents walked for recreation. Those in areas ranked highest on 'perceived greenness' were 44 per cent more likely to have good mental health than those in areas of low perceived greenness and this difference was statistically significant.

Other questionnaire based studies aim to improve our understanding of *why* people enjoy natural landscapes. In a study exploring the perceived therapeutic benefits of natural settings, Palka (1999) conducted a questionnaire among visitors to Denali National Park, Alaska, distinguished by its vast extent, devoid of human settlement and its diversity of flora and fauna (Figure 2.4). Visitors have access to the park via bus tours along a 'corridor', interrupted by rest stops. The area otherwise has few trails. The attributes of the landscape that the visitors appreciated included: the *pristine*, unspoilt, unpolluted natural landscape; its aesthetic appeal due to its impressive *scale, beauty*, and *diversity*; *wildness* of the setting and *abundance* of wildlife; the *remoteness* and *quietness* of the area offering a sense of undisturbed *solitude* and *escape* from more stressful, manufactured and commercial landscapes. Informants talked about feeling more relaxed and revitalised. Palka's study found that these elements of natural scenes

Figure 2.4   View of Denali national park

Note: The grandiose scenery of Denali National Park Alaska, was investigated by Palka (1999) to explore how visitors responded to its 'therapeutic' properties

Source: NPS photo by Kent Miller, http://www.nps.gov/dena/naturescience/index.htm

also seemed to be valued in terms of their symbolic and spiritual significance as spaces which are 'authentic' and 'unsullied' by baser human behaviours. Contact with natural landscapes also increases opportunities for interactions with the animal kingdom. A review of geographies of human–animal relations edited by Philo and Wilbert (2000) includes a number of examples of research which explores the relationships between people and animals and how these influence our experience of 'green' and 'blue' landscapes.

The examples considered above illustrate empirical research which seems to demonstrate health benefits associated with access to natural, green spaces, consistent with some of the theories reviewed so far in this chapter. Further research has explored how particular aspects of these spaces seem to relate to mental state. In order to enquire further into the health effects of 'green' and 'blue' landscapes postulated by De Vries et al. (2003), the following discussion focuses on selected studies which have explored the significance for mental health and wellbeing of trees and woodland, and the psychological response to waterscapes.

The importance for mental health of woodlands and trees

A growing literature from a range of cultural and social settings supports the idea that trees and woodland are beneficial to psychological health and wellbeing. The examples considered below are partly based on the kinds of arguments put forward by theories of 'biophilia' and 'restoration', concerning the ways that natural landscapes may satisfy our 'deep psychology' based on 'animal' instincts for physical survival, and offer respite from the stresses of 'unnatural' surroundings. However, they also invoke human responses which seem to draw on spiritual and ethical values and the importance of social relationships.

In Japan, for example, health benefits have been attributed to *Shinrin-yoku* (walking and/or staying in forests and 'forest air bathing' – inhalation of the vapours transpired by forest trees). An experiment involving 12 volunteers, recording physiological measurements for blood pressure and levels of the 'stress related' hormone cortisol, is reported by Park et al. (2008). They found that physiological stress levels on these measures were lower after practising *Shinrin-yoku*. Morita et al. (2007) studied a sample of about 500 volunteers while they were visiting woodlands for *Shinrin-yoku* on two different days, and, for comparison, on two other days when they were not visiting the forest. The researchers collected responses on questionnaires on self reported mental state, as well as other information including socio-demographic characteristics and whether or not the respondent enjoyed visiting woodland. They found that stress levels (especially feelings of hostility and depression) were lower (better) on the 'forest days' than on the 'control days' (see Figure 2.5). The differences were more pronounced for people whose psychological health over the preceding two months had been relatively poor, and also for women and younger people. A large majority of the volunteers enjoyed forest landscapes and were likely to be able to engage in their favourite pursuits there, and on 'forest days' they were likely to be physically active, which may have benefited their mental state. However, even people who did not especially enjoy these woodland settings and were physically inactive on 'forest days' appeared less stressed while they were in the woods. The authors argue that experience of these natural woodland settings has a therapeutic effect. According to Morita and colleagues, the Japanese Ministry of Health, Welfare and Labour encourages regular interaction with natural environments as part of their health promotion strategy and the research findings seem to provide evidence in support of this.

The idea of natural settings as spiritually and morally uplifting is also reported in geographical research from Britain which explores the psychological benefits of social forestry projects. Bell and Evans examined discourses about benefits of natural landscapes presented in plans for the creation of the 'Millennium Forest' a major urban forestry project undertaken to mark the millennium and involving planting and management of publicly accessible urban woodlands across the Midlands in England, predominantly through reclamation of disused urban-industrial land (Bell and Evans, 1997; Bell, 1999). Their analysis of the rhetoric in public consultation, planning documents and planning enquiries identifies the

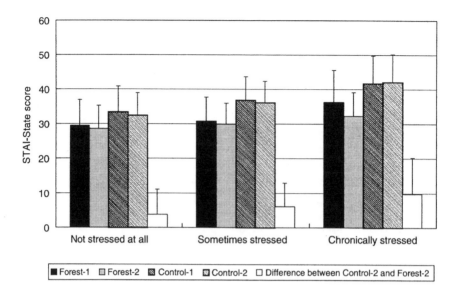

Figure 2.5   *Shinrin-yoku* and psychological stress levels

Note: Mean scores on a measure of psychological stress for three groups of study participants who were generally not stressed at all, sometimes stressed or chronically stressed. Current stress levels were measured on two forest days, spent practicing *Shinrin-yoku*, and on two control days. Stress levels were lower on the 'forest days' than on the control days', especially for people who reported they were 'sometimes' stressed or chronically stressed over the last two months.

Source: Morita et al. (2007: 60). Reprinted from *The Lancet* by permission of Elsevier.

idea of the 'redemptive' symbolism of the project, partly because of the restoration of the land to more sustainable uses and also because of the social benefits from the cooperative and participative aspects of the projects involved in creating community forests and providing improved to green space for all members of the public. Their account suggests that woodland landscapes are seen to symbolise fundamental moral values such as integrity and purity, and environmental justice and social inclusion.

Other research suggests, however, that these human responses to natural settings such as woodland are not universal. Milligan and Bingley (2007) explored the evidence for restorative effects of woodland spaces for young adults interviewed in northern England. Some of the participants made comments that seemed to support the idea of woodland as 'restorative' and therapeutic. Comments from the participants suggested that time spent in woodland restores psychological balance, conveys a reassuring sense of stability and continuity, and offers an environment which is relatively 'pure'. For example, Milligan and Bingley cite comments including the following extracts (abbreviated here from their original report):

It clears your mind, I think...You're on your own and it's peaceful...it flows though your mind and everything goes back into perspective...

...these trees...they've been here for hundreds of years...it's...kind of steady

The air is ..cleaner and lighter...fresher... (Informant's comments cited by Milligan and Bingley, 2007: 804–5)

However, there were variations in responses and other informants in this study were more ambivalent or felt fearful or repelled by woodland (see Chapter 3). Some wooded landscapes appear threatening and repellent to some people, especially dense, impenetrable woodland where unseen dangers may be lurking. Figure 2.6 shows two different types of tree filled landscapes which might therefore be therapeutic in different ways to different people. The urban woodland scene may be attractive to some people as it illustrates accessible but rather natural looking trees, while the other corresponds to the more cultivated and managed 'arcadian' parkland vision of trees which seems preferable to other groups of people.

Other studies also suggest that it would be misleading to assume that human responses to natural environments are universally similar; they emphasise the differences in perception and preference between social groups. Research investigating adults responses to 'virtual' representations of environment, is reported by De Groot and Van den Born (2003). They found that people in their sample from the Netherlands distinguished between different visions of what comprised 'natural' settings and varying aspects were preferred by different social groups of respondents. Those with lower levels of educational attainment tended to consider 'cultivated' and controlled aspects of 'arcadian' nature captured in agricultural scenery (e.g. ideas of lambs in the meadow, grain fields, pollarded willows). 'Elemental' nature recalling wilder natural elements (e.g sea, wind) were preferred by more educated groups in their sample. Respondents who preferred more elemental nature also were more likely to assess as 'natural' the descriptions of 'elemental' images or ideas of 'penetrative' nature ('mosquitoes', rats in the barn', 'weeds in the garden'). Those preferring arcadian nature were more likely to assess more cultivated settings and parkland as 'natural'. Thus social and cultural experience may influence views of natural settings. Furthermore, Herzog et al. (2002) reported that a student sample in the USA found activities in more 'artificial' and 'urban' settings such as cinemas and nightclubs more attractive than natural settings.

These studies of the mental health benefits of landscapes including trees and woodland therefore combine to give a multifaceted view. While several studies seem to offer empirical evidence in support of the idea that trees and woods have

**Figure 2.6   Varying appeal of different visions of trees: different
environments including trees**

Note: Woodland spaces can provide access to tree filled natural landscapes, which research
shows are beneficial to mental health for some people. 'Urban woodland' illustrated in these
pictures taken near the centre of Durham, and in Greenwich, London, UK, offer access to
many city dwellers to the 'restorative 'aspects of woodland settings.

therapeutic properties for mental health, these are clearly variable among different
social groups and from one individual to another and green settings are not the
only type of landscape that people find attractive.

Water as a landscape element related to mental health

As we have seen, some research aims to consider human responses to specific landscape elements to try to identify the crucial aspects of 'green' and 'natural' environments for people's psychological state. In addition to research focused on trees and vegetation discussed above, many studies consider water features in landscapes as having special symbolic and social significance for health. All of the theoretical frameworks considered at the start of this chapter suggest that certain types of waterscape are likely to have therapeutic properties.

Water is an essential element for life, and has mythical and religious significance as medium of purity and cleanliness in a moral, as well as a physical sense. For many places with a reputation for healing, water is a crucial element of the landscape and this is especially evident for spa towns. Geographical researchers using a therapeutic landscapes framework have examined spa towns, including Gesler (1998a) in his interpretation of Bath spa in south west England.

Figure 2.7 illustrates different parts of the natural, hot springs which feed Bath spa, and which have attracted human activity for thousands of years. The spa waters at Bath are a good illustration of the role of therapeutic natural water and a key factor in the healthy reputation of a healing place. They were recognised in myths and legends by ancient Britons, before the Roman invasion, after which, in 43AD, the site was dedicated to roman deities and was constructed as the centre of a major Roman health and leisure complex, attracting visitors from many parts of Europe included in the Roman empire of the time. Subsequent revivals of the site included its development as a site of pilgrimage under the patronage of the Christian church, during the medieval period, and as a fashionable leisure resort, which boomed in the mid-18th century in response to a new enthusiasm for 'taking the waters' in spas. The spa facility at Bath was again refurbished and updated at the turn of the millennium as the *Thermae Spa* with a modern commercial promotion of the place as a health spa, which clearly draws on the reputation and historical significance of the hot springs.

A similar link to economic consumption and production as well as the search for settings with health-giving, restorative properties, is emphasised in Goeres' (1998) account of Hot Springs, South Dakota, USA and its promotion as a spa town in the early 1900s. The following extract is from the text of publicity material advertising the resort, cited by Goeres, that emphasises the importance of water as a key attribute and incorporates ideas about water in the springs as a 'comfortable' medium, imbued with 'natural' curative properties, and characterised by wholesome properties such as purity:

> There is no other place in the United States which combines the three great essentials of comfort and health in so perfect a manner as they are combined at Hot Springs...waters tempered to man's normal heat, charged with Nature's best remedies and given out in great purity and abundance. (From a publicity pamphlet by Northwestern Line (1910), cited by Goeres (1998: 45); abbreviated from the original quote).

**Figure 2.7a   Scenes of Bath Spa (SW England): The ancient monument of
the Roman Baths of Aquae Sulis dating from around 43AD**

Source: http://www.thermaebathspa.com/visitorcentre/spa_history/to-1090/index.html.
Reprinted with permission of Roman Baths, Bath & Somerset Council.

**Figure 2.7b   The new Thermae baths offering modern health spa luxury,
still centred on the natural hot springs**

Note: These spa developments, ancient and modern draw on the idea of the healing
properties of water, both for mental and physical health.

Source: Cardy, M.(2009) Minerva bath by day, Thermae Bath Spa, reprinted with permission.

Spa towns and centres of pilgrimage have always focused as much on psychological and spiritual health as on physical health, and questions concerning the spiritual significance of landscapes are taken up in more detail later in this book. For the present discussion we are interested in how the perceived therapeutic and 'natural' properties of water contribute to psychological health and what makes water a key physical and symbolic element of therapeutic landscapes. Spa towns rely for their success on the relationship they establish between 'healing' water and a range of 'restorative' and engaging human activities that spa towns aim to make possible through the facilities they provide for retreat, indulgence, and cultural and social pursuits. These aspects underline interpretations of water as a benign element for human wellbeing, due partly to 'deep psychology', predicated by biophilia and restoration theories. Research on therapeutic landscapes also suggests that the response to waterscapes is socially constructed and contingent on cultural factors that construct the social and symbolic significance of water for health in varying ways.

Environmental psychologists have conducted a number of studies attempting to analyse the various attributes of water and waterscapes that people find attractive, satisfying and restorative. Their work draws particularly on theories concerned with biophilia and restoration. Examples include: Moser (1984); Herzog (1985); Herzog, and Bosley (1992); Herzog and Barnes (1999); Hetherington (1993); Wilson, et al. (1995). Selected examples of findings from these studies are summarised in Figure 2.8. In these examples, we can see that the groups of people studied are rather specific, comprising students and others working in American universities; such groups of informants predominate in this literature, since academic psychologists working in this field are not especially concerned with social variation in perceptions and they have often recruited 'convenience' samples from among their students or colleagues. In some respects, this narrow focus on particular types of people makes the findings all the more interesting, since we see a range of different attributes of water identified by rather similar groups of people.

Aspects of water that were attractive to participants in these studies included: extensive waterbodies such as mountain lakes and seascapes, favoured for the sense of space and tranquillity they conveyed; rushing water in streams was also attractive, especially for the sound as well as the visual sense of movement; clear, clean water, as opposed to polluted water bodies; waterscapes offering opportunities for particular activities such as water sports. The variety of attributes of water that were identified probably reflects differences in the ways that participants were asked to evaluate water scenes (e.g. in terms of preference or in terms of certain impressions afforded by the landscape such as tranquillity or fascination). It is also clear, even from this limited set of studies that perception of these waterscapes is influenced by the other landscape elements in or around the water, by the kinds of engagement with the landscape that people are able to envisage, and by the sorts of activity they can imagine they might do there. Wilson et al. (1995) found that edited photographs of open water, altered to

Author	Study Design	Findings
Moser (1984)	85 interviews with holiday makers in France	water quality judged in terms of debris; colour, odour and water plants
Herzog (1985)	250 psychology students, USA	preference was for spaciousness, coherence and mystery, and for mountain lakes and streams
Herzog (1992)	341 undergraduate students, USA	large bodies of water were seen as more tranquil but rushing water was more likely to be preferred
Hetherington et al. (1993)	339 university students, USA	respondents preferred scenes with faster flowing water which were accompanied by sound
Wilson et al. (1995)	105 people from university community, Canada	respondents preferred scenes showing canoeists and a goose. These were preferred for recreational activity. They disliked scenes with a floating tyre, aquatic vegetation; surface foam; a health warning sign; an industrial backdrop

Figure 2.8   Selected examples of research in environmental psychology on the attractive elements of waterscapes

include different elements such as waterbirds, floating tyres and debris, or people kayaking were assessed differently in terms of the quality and attractiveness of the water. Thus we see evidence of an interaction between human motivations and agency and the psychological response to water. The pictures in Figure 2.9 illustrate some of the features of waterscapes which are referred to in these studies and readers may like to think about whether these, or different images of water scenes would be attractive to them and what makes such images seem appealing. Many of these studies rely on still photographs, which are selective representations of real landscapes and may not capture the full impression to be gained from direct experience of the real setting. Hetherington (1993) showed that moving video images supplemented by sound were evaluated differently than still photographs.

As for the woodland scenes discussed in the previous section, we see that the material attributes of water, the accompanying landscape features, and the types of human activity possible in waterscapes all contribute to the perception of therapeutic landscapes which are benign for mental health and wellbeing. Approaches which analyse landscape in terms of the specific elements which seem attractive to people demonstrate that complex combinations of elements, rather

Figure 2.9   Some aspects of water that people seem to find attractive

Note: Pictures a), b) and c) respectively illustrate: a) waterscapes offering opportunities for enjoyable activities such as rowing, b) fast flowing streams and c) waterscapes including appealing wildlife.

than particular features viewed in isolation, seem more important, and we see from some of these studies that varying value systems and socially constructed frames of reference seem to influence perception as well as the 'deep', genetically encoded psychological responses postulated within theories of biophilia. Furthermore, human relationships with nature are not all harmonious and benign, and as discussed in Chapter 3, natural landscapes and natural events in which water is experienced as hazardous or threatening can also produce more negative psychological responses.

This observation concerning the complexity of our response to nature supports the perspective on therapeutic landscapes invoked by authors such as Gesler (2003) in his view of 'therapeutic landscapes'. As was also discussed here in Chapter 1, Gesler suggests that 'therapeutic landscapes' need to be understood in terms of interacting 'fields' of physical (material), social and symbolic aspects of landscape. Also, natural and fabricated attributes of the material environment are included in Gesler's model as combining together in therapeutic landscapes, so the following section, considers material attributes of 'built' landscapes that may benefit mental health.

Urban landscapes and built forms

To an extent, the conceptual frameworks considered above, such as biophilia and restoration theory, can be applied to the assessment of urban settings, particularly in terms of the extent to which they incorporate natural elements that may benefit health. However, Gesler's (2003) therapeutic landscapes framework suggests that the built and fabricated elements of material landscapes are also important for our wellbeing, so we need to draw on additional theories are needed to interpret the ways that people respond psychologically to built structures.

One perspective that is useful in this respect is the idea of *Topophilia*, put forward by Yi Fu Tuan, which he defines as including "all of the human being's affective ties with the material environment" (Tuan, 1974: 93). It provides a conceptual framework for considering the emotional bond between people and places. This emotional response to place is influenced by one's cultural frame of reference and is linked to sense of individual and group identity. Many of the examples discussed by Tuan (1974) concern the social and symbolic significance of the design of buildings and settlements, as well as topographical environments of 'constant appeal' – seashores, valleys and islands. Tuan's work places considerable emphasis on culturally specific interpretations of landscape, and this raises important questions about how far human responses to environment are universal and based on biological and genetic imperatives. For example, Tuan (1974, Chapter 8) proposes a model of changes through time in human perceptions of 'edenic', sacred, ideal landscapes and 'profane', threatening and repellent settings. In ancient European cultures urban settings were often represented as ideal and desirable sites for human existence. For example, Plato's *Republic*, written in Greece in 360BC, presents a social, political and moral programme for civilised human society which is envisaged as located in a city not in a rural setting (see translation by Jowett, 2009). Garden settings, typifying Arcadian landscapes were also, for much of human history idealised and contrasted with 'profane' wilderness settings (e.g. the biblical 'Garden of Eden', American utopian communities of the early 19th century and the Garden City movement in England. In some cases, nature was also seen as 'edenic' (usually represented as Arcadian rather than wilderness landscape) and was viewed as complementing more sophisticated, 'cosmic' urban society. For example, 18th and 19th century landscape gardens in England were being developed at the same time as urban construction of buildings in the 'classic' style, favoured by enlightened intellectual perspectives that drew on ancient European knowledge and philosophy. Bath spa, discussed above, is famous for its magnificent 'classical' architecture of the period (see below, Figure 2.12). Tuan's model suggests that, at least in western societies, wilderness landscapes only began to be represented as ideal and edenic during the industrial revolution of the 19th century, which generated massive and noxious urban development. During the aftermath, in the contemporary period, we have become more concerned for conservation of 'threatened' natural wilderness. Such natural landscape is now seen as offering an escape from the 'new urban wilderness' of impoverished, crowded and polluted city centres and sprawling, soulless suburban development.

Tuan's arguments about *Topophilia* therefore lead us to consider varying intellectual and cultural, as well as biological responses when investigating the links between material landscapes and mental state, and he encourages us to be sensitive to the culturally constructed and historically variable character of landscapes which are viewed as beneficial for human wellbeing. For example, to those whose cultural frame of reference is based in 'western' (originally European) cultures, ancient Chinese theories of *Feng-Shui* offer an exotic and intriguing perspective. Ideas from these relatively sophisticated intellectual schema for analysing the therapeutic properties of built landscapes, have been widely adopted and adapted in western culture. As explained by Chiou and Krishnamurti (1997), *Feng-Shui* expresses ideas of how the disposition of elements in space and time help to ensure a healthy accumulation of *Qi*, the life enhancing energy which is a key component of health according to Chinese medicine.

Feng-Shui is of relevance for the discussion here because of the ways that it interprets the physical placement of material things in relation to culturally constructed ideas about harmony and wellbeing. These schema have a very long tradition in Chinese culture for design of physical settings believed to be beneficial for wellbeing. Feng-Shui is known to have been applied as early as 276BC to the choice of location for graveyards and in the 11th century to house design. Knapp (1999) has studied the influence of these beliefs on home design and ornamentation, and Mak and Ng (2005) comment that the 'form' school would be likely to identify successful locations for settlements which would be protected from natural hazards.

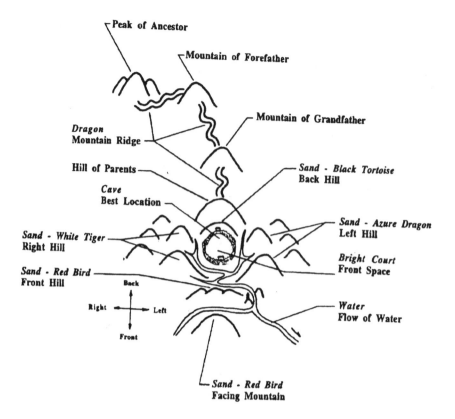

Figure 2.10   Representation of the *Feng-Shui* model

Source: Mak and Ng (2005) Figure 2. Reprinted from *The Lancet* with permission of Elsevier.

Elements of Feng-Shui

According to the *Feng-Shui* school of 'earthly forms', *Qi* is carried by winds or held by water so a harmonious setting is created by an auspicious placement of topographical features to accumulate Qi, as shown in Figure 2.10. The disposition of features such as mountains, sandhills and water around the 'cave', or home site, was thought to be required for these harmonious conditions that 'store' Qi. Chiou and Krishnamurti (1997) and Mak and Ng (2005) discuss the distinction between this 'form' school of *Feng-Shui* and the 'compass' school. The latter relates to cosmological and astrological readings of particular locations, based on a form of geomancy using the model of the *Feng-Shui* 'compass (illustrated in Figure 2.11). This combined ideas about periods in the year, fundamental elements making up the physical world, and astrological readings which would vary for each person depending on factors such as their date of birth and sex. Thus, according to *Feng-*

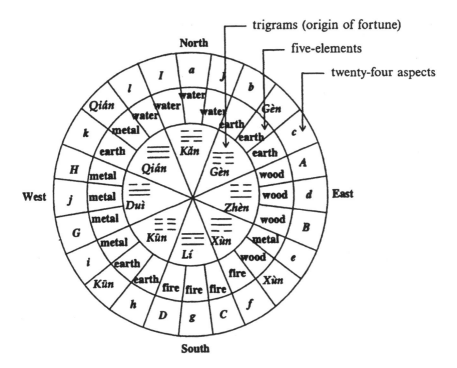

Figure 2.11   The *Feng-Shui* Compass

Source: Chiou and Krishnamurti (1997). Reprinted with permission of Pion Ltd, London.

Shui, it is possible to identify, for each individual the ideal placement of things and an auspicious time at which to commence a project such as construction of a building.

As western cultures have been increasingly influenced by ideas from Chinese culture (including Chinese traditional medicines involving therapies using herbal remedies, acupuncture or Tai Chi exercises, for example), we have also seen a growing interest in *Feng-Shui* as a basis for healthy design. In Chapter 6 there is further discussion of the symbolic significance of landscape bound up in schemes like *Feng-Shui*. Here we note that journals of architecture and planning include a number of references to *Feng-Shui* and ways it may be applied to built forms. For example, there have been proposals e.g. reported by Lind (2004) that *Feng-Shui* should be promoted in areas of California. Hwangbo (1999) comments that the growth of interest in this perspective may be motivated by the need to assuage a sense of nihilism in modern western society, as well as an attraction to the exotic. This illustrates globalisation of cultural understandings of therapeutic landscapes (a theme that is taken up again in Chapter 6).

'Architectures of happiness'

A wider view of the relationships of built structures to human wellbeing is presented by Alain De Botton (2006), who demonstrates the huge diversity of different built forms, interior decorations and styles of landscape architecture that have appealed to human aesthetic, emotional and physical senses over the centuries. While he acknowledges the significance of the attraction of forms which recall and incorporate natural elements, he also presents an argument about the aesthetic 'virtues' of buildings that we find attractive (De Botton: Chapter V, pp. 169–249). These include: *order*; *balance*; *elegance*; *coherence*, as exemplified, for example, in the superb Georgian architecture of Bath Spa (Figure 2.12). A further element of De Botton's scheme is *self knowledge*, and he highlights efforts to use our knowledge of problems in human societies to improve the human condition, through modern reconstruction of urban fabric of historic settlements. However, De Botton makes the point that in this respect we have often failed due to the destruction of other elements that we value, such as ideas of heritage, human scale and diversity. His critique of the impracticability of Le Corbusier's ultimately 'de-humanising' visions for urban form provides a case in point. De Botton's analysis therefore sounds a warning as well as offering ideas about how to promote 'happiness' and wellbeing through architectural design.

It has almost become a cliché to discuss historic buildings, such as the Royal Crescent in Bath, in terms of their uplifting and attractive character and their special value as part of our cultural heritage. Such interpretations emphasise the social and cultural significance of these structures as much as the intrinsic material features of the design, so that to fully understand the impact that these buildings have on our imagination we need to consider social and symbolic attributes of therapeutic environments, as discussed in later chapters of this book. Newer buildings in a neo-classical style make reference to similar values, though they often combine an eclectic mix, using diverse stylistic features and modern materials to invoke the 'virtues' that de Botton identified. A striking example is the *Antigone* in Montpellier, France (Figure 2.12), which conveys, in a modern cement construction, many allusions to classical architecture, and also includes several of the attractive natural elements of trees and water that were discussed above. It is interesting to interpret the *Antigone* in light of theories concerning Topophilia and architectures of happiness.

This example also provides a preview of some of the themes in Chapter 4 of this book concerned with social relationships. Families from the housing around this square, gather and mix socially using the café tables around the fountain together with visitors to this tourist town. There is no motorised traffic and the space is devoted to relaxation and leisure. The scene combines the physical landscape elements, the symbolic references to classical enlightenment philosophy and an inclusive social setting that combine to create what might then be seen as an ideal therapeutic landscape. It is accessible, safe, predictable, and civilised. Many

Figure 2.12   The Royal Crescent at Bath Spa, England, an example of the virtues of architecture put forward by De Botton (2006)

Note: Buildings with regular, repeated forms and well organised form may appeal to our need for a reassuring sense of *order* and control of our environment, and our preference for *coherence* in architectural forms, as exemplified, for example by the perfect rhythmic structure of the Crescent in Bath. The classic architectural style of the Crescent also offers perfectly proportioned and *balanced* elements in the facade, constructed according to Palladian principles, which are also rendered more 'elegant' by the smaller stylistic flourishes and the curved structure of the building which helps to prevent a sense of 'heaviness' or dullness. Possibly this row of houses also appeals to ideas of *self knowledge* in the sense that it reflected the 'rationality' of enlightened thinking which, during the 18th century in Europe, was associated with rapid strides in philosophical, scientific and artistic endeavours in society.

Source: http://visitbath.co.uk/site/home. Reproduced with permission.

people prefer to experience this type of artificial and managed environment, rather than in a pristine, wilderness setting.

Individual scholars may attribute significance to built form in these ways, but we cannot establish how far these views may be widespread and consistent in the general population without empirical research to explore how wellbeing and good mental health are related to 'health promoting' material features of urban environments. Some empirical research uses rather systematic, extensive questionnaire techniques to investigate a range of potentially therapeutic attributes of built environments as they relate to human wellbeing, drawing on more than one theoretical perspective. For example, Obunseitan (2005) elides ideas from Topophilia and Attention Restoration Theory in a study of perceptions of a sample recruited on an American University Campus. Respondents were asked to assess their current home environment in terms of questionnaire items relating to different dimensions: *cognitive* aspects (mystery, familiarity, coherence), *synthetic* tendency (aspects appealing to the senses such as colour, light, sound), *familiarity* (identifiability and privacy) and *ecodiversity* (flowers, trees, water, topography). They were asked, as well, for their assessment of their home space in terms of its restorative properties (the question asked about 'the abundance and variety of restorative elements that are accessible to you'). Their ratings of

**Figure 2.13   Scene from the Antigone, Montpellier, France: an example of
a therapeutic landscape in an urban setting**

Note: The Antigone is an urban square designed by Ricardo Bofill and built in the 1980s in
the centre of the ancient and historic city of Montpellier in Southern France. This picture
of the Antigone could be interpreted in terms some of the key themes of therapeutic
landscapes, drawing on theories such as biophilia and topophilia. Water has an appeal for
humans which may be partly based in deep psychology and individual preferences but is
also culturally and socially constructed. Water is well under control here in an urban setting
and the children are enjoying it in the hot weather. The fountains are intermittent, which
provides extra interest and a sense of 'mystery' as the jets of water rise and fall in ways
that are hard to predict. There are trees providing shade and greenery, but not obscuring
the view, corresponding to the 'Arcadian' sense of a 'controlled' and well regulated natural
environment. The architecture plays a key role too. It draws on classical architectural themes
– the curved forms, statues and columns echo the Greek, Roman or renaissance Italian style
which offers coherence and order and balance in architectural design, as well as invoking
symbolically the Mediterranean heritage of this historic southern French town.

their home environment on the different dimensions were found to be associated
significantly and positively with their rating of the restorative properties of their
home. Obunseitan (2005) also reports a significant statistical association between
the ecodiversity ratings and respondent's reported quality of life and on the
WHOQOL-Bref quality of life measure (discussed in Chapter 8). This finding

suggests that natural environmental elements were more important for human wellbeing than built form. It is further suggested that this study illustrates a way to apply ideas from theories such as Topophilia to assessment of urban and domestic design. However, the approach does have limitations. For example, the study collected data on the participants' age, sex, education levels and ethnicity, and the author reports that the sample was relatively young and highly educated, but these socio-demographic variables were apparently not controlled for in the statistical analyses. Also, since the information was entirely based on respondents' perceptions and is cross sectional, we cannot tell from this study whether there is any direct causal association between perception of the home environment and quality of life and in which direction causality might operate.

This example, therefore, also serves to highlight the scope to assess these issues in more depth, through research using some of the other strategies discussed later in this book, such as qualitative research to explore in detail how and why certain material surroundings produce positive psychological responses and by paying greater attention to social and cultural factors. There may also be scope to conduct experiments to assess whether experience of 'therapeutic' material environments seems to influence subsequent health and whether by creating settings with more therapeutic qualities public mental health may be improved.

Combining theories for a complex view of therapeutic landscapes

The discussion above has emphasised the continuing relevance of the following statement by Penning-Rowsell (Penning-Rowsell and Lowenthal 1986: 117), who argued that:

> Preferences and prejudices could derive from a hierarchy of elements: biological characteristics, social systems, personal characteristics...The question remains: how much in landscape taste is universal and how much diversity is there?

This question still seems very relevant today, and it suggests that a more comprehensive and possibly more 'serviceable' view of how material elements of built environments can be beneficial to mental wellbeing should to adopt a complex perspective which combines elements of all of the theories discussed above. Drawing on Biophilia, Restoration Theory and Topophilia we gain a more complete understanding of psychological responses to material landscapes, resulting from interactions between the material environment, one's genetic make-up, individual personality and socio-cultural frames of reference. This agenda also calls for a mixed methods approach to research. Some of the examples reviewed in this chapter have used quantitative research methods effectively to study patterns of response to the environment, while other studies deploy qualitative methods. While a good deal has been learnt from analytic strategies which break down the landscape into specific material components that are important for psychological wellbeing, a more holistic and synthetic strategy will also be necessary to understand better how these elements combine and interact.

Referring back to the discussion of phenomenology in Chapter 1 of, we may also consider Porteous' (1990: 87–105) discussion of the idea of *inscape*, which expresses the ways we psychologically internalise aspects of landscape, producing a 'blurring of the landscape and the mind'. Porteous argues for a 'participatory' role in the man-nature relationship that may be more sane than the current 'mastery' world view of human interactions with nature...breaking down the "alienation from both nature and our own being that is characteristic of the human condition in western society". We return to this theme in Chapter 8 where there is discussion of human psychological responses to global environmental processes and links between psychological health and sustainability.

Key learning objectives

Key knowledge to be gained from this chapter relates to:

- theories which inform our understanding of the relationships between mental health, psychological healing and aspects of the material environment (especially Biophilia, Topophilia and 'Attention Restoration Theory')
- the significance of 'green' and 'blue' landscapes for mental health;
- the characteristics of 'architectures of happiness'

Introductory reading

Reading on the idea of therapeutic landscapes includes:
Gesler, W. (2003). *Healing Places.* Lanham, MD: Rowman and Littlefield.
Williams, A. (ed.) (1998). *Therapeutic Landscapes: The Dynamic between Place and Wellness.* Boston: University Press.
Williams, A. (ed.) (2007). *Therapeutic Landscapes.* Aldershot: Ashgate.

For discussions of research on ideas such as Biophilia:
Kahn, P.H. (1997). Developmental psychology and the biophilia hypothesis: Children's affiliation with nature. *Developmental Review*, 17(1): 1–61.
Kahn, P., Severson, R. et al. (2009). The human relation with nature and technological nature. *Current Directions in Psychological Science*, 18(1): 37–42.

For the original presentation of Topophilia:
Tuan, Y.-F. (1974). *Topophilia: A Study of Environmental Perceptions, Attitudes and Values.* Englewood Cliffs, NJ: Prentice Hall.

Also, for an accessible discussion of related ideas concerning 'architectures of happiness':

De Botton, A. (2006). *The Architecture of Happiness*. London: Penguin Books.

Chapter 3
Material Hazards and Risks for Mental Health

Summary

The previous chapter examined aspects of the physical landscape which may benefit mental health. The discussion here considers the converse of therapeutic landscapes; hazardous environments, which constitute harmful risks for mental illness or psychological distress. In part, such risks are created by depriving people of some of the benign landscape elements reviewed earlier. Also, features of the environment which benefit some people's health are harmful for others. However, this chapter mainly focuses on some aspects of the material world that are more actively damaging for mental health, including the associations between mental state and natural and manufactured hazards, disaster events and deprived material living conditions, which also contribute to socio-geographic variation in mental health.

Introduction: Natural hazards and mental health

The natural material elements discussed in Chapter 2 as having benign effects may, under different conditions, be construed as hazardous or threatening. These negative reactions may arise because of individually and socially variable experiences and frames of reference. For example, in the study which we considered above, investigating health benefits of urban woodland, Milligan and Bingley (2007: 806) also reported that some participants expressed fear and dislike of woodland settings. 'Scary' aspects included "...The darkness...what's hiding in there, crawling in there", feeling "trapped" or threatened by "...anyone lurking" in the vegetation. Some respondents also thought of woodland as dirty. Fears such as these were found to be associated with early influence of parental attitudes about woodland as suitable play space for children, and the influence of media and myths relating to woodland settings, showing how perceptions of natural environments are socially constructed and contingent on personal experience. Similarly De Groot, and Van den Born (2003) found that for some respondents, especially those who preferred more 'tame' or 'cultivated', Arcadian views of nature, certain aspects of natural environments were repellent or unattractive, typified by ideas of 'penetrative' nature.

In other circumstances, however, it is the natural elements themselves which vary in time or place and may become hazardous. This is most dramatically evidenced in the case of 'natural disasters'. The World Health Organization (1992: 2) defines a disaster as: 'a severe disruption, ecological and psychosocial, which greatly exceeds the coping capacity of the affected community'.

Some such disruptions are viewed as 'natural disasters' (although human activities often influence the risks and the outcomes of these events). They include meteorological events, such as droughts and extreme heat, or blizzards, hurricanes and tornadoes, or geological events such as volcanoes, earthquakes and landslides. Some of these may be combined with hydrological disasters such as floods and tsunamis. Epidemics of disease sometimes are associated with processes in the natural world and, in severe cases, may be viewed as 'natural disasters'. Other disaster events are primarily generated by human actions. They include non-intentional spread of infectious diseases due to human activity, episodes of environmental contamination or other industrial accidents, transportation accidents, and intentional violence (terrorist attacks and military conflicts).

Reviews of research on mental health of people involved in disasters (e.g. Norris et al., 2002a, 2002b; Norris, 2005; Galea, Nandi, and Vlahov, 2005; Ahern, Kovats et al., 2005, 2006; Neria et al., 2008; Curtis and Nadiruzzama, 2009) have summarised evidence that experience of disasters is associated with various mental health problems. The most commonly reported disorder is post-traumatic stress disorder, and other conditions include non-specific distress, somatic symptoms, anxiety, depression, as well as related behavioural changes (e.g. increased drug/alcohol use; suicidal thoughts and actions).

Some accounts of psychological impacts discussed below also report instances of more positive outcomes from traumatic experience of disaster, sometimes typified as 'post-traumatic growth' (PTG). PTG describes the way that some survivors are able to learn from their experiences, and find new psychological strength in the knowledge that they have been able to survive the disaster, giving them a stronger sense of self efficacy, empowerment and resilience to adversity.

Post Traumatic Stress Disorder (PTSD) is defined as a psychological 'crisis after a stressful event of exceptionally threatening or catastrophic nature' and can also be thought of as 'traumatic memory'. Typical symptoms may include: 'intrusion' ('flash backs'); 'avoidance' (distancing oneself from the event); 'hyperarousal' (eg. irritability, difficulty concentrating, hypervigilance). Not all experts agree that PTSD is a psychiatric 'disorder' (Ramsay, 1990; Bracken et al., 1995; Young, 1995). For example, Summerfield (2001) suggests PTSD is better understood as a 'social condition' that has been redefined as a 'clinical disorder'. He argues that the disorder is a relatively recent invention by psychiatrists, first recognised in association with legal proceedings over benefits claims of American Vietnam veterans who were unlikely to win claims unless a medical condition could be attributed to their experience of the conflict. It is also suggested that the PTSD diagnosis is over-simplified, attributing the symptoms to a 'single cause' (experience of the disaster) rather than complex causal factors. Also, it may be seen as based on individualistic western ideas about the importance of the self, so that it may be 'culture bound' and socially constructed, possibly associated with the anxieties generated by conditions of modern life (see discussion below and in Chapter 8 relating to Beck's (1992, 1999) theories of the 'risk society'). According to Summerfield it is difficult to distinguish PTSD from 'normal' distress provoked by traumatic events. Summerfield's arguments are interesting in relation to the discussion in Chapter 1 and later in this book about varying perceptions of mental health and illness. From an anthropological perspective Kienzler (2008: 222) reminds us that "suffering is resolved in a social context and familial, socio-cultural, religious and economic activities make the world comprehensible for people before, during and after catastrophies", and that it is probably not correct to assume that all individuals will be affected by disasters in the same way, or will benefit from the same types of healing strategies.

However, there is apparently a widely held view among psychiatrists, that PTSD as a medical diagnosis has clinical validity. Other authors have reviewed a long history of accounts dating back to the 19th century of PTSD symptoms (Keinzler, 2008) and the symptoms are widely reported in studies in many different countries (including those reviewed below). Mezey and Robbins (2001) take the view, for example, that PTSD transcends 'normal' misery/unhappiness, and is associated with significant physiological changes (those affected by PTSD have increased levels of the stress related hormones cortisol and corticotrophin, and the part of brain involved in memory becomes hyperactive). Such physiological changes in the brain and the nervous system are also interpreted by Bracha (2006), in terms of genetically determined, biophysical responses to fear and stress, emphasising neuro-evolutionary theories of brain function. PTSD is also argued to be important because it increases risk of other disorders like depression and social problems like unemployment and marriage instability.

Survey questionnaire instruments to record symptoms of PTSD have been developed for use in population studies investigating the health effects of trauma. (For example the Impact of Events Scale is illustrated in Box 3.1) Research based on national sample surveys at different time points have made varying estimates of the 'background' prevalence of PTSD, suggesting, for example, that the condition may typically affect between 1 per cent and 8 per cent of the general population in America (Mezey and Robbins 2001). This 'background' prevalence could be due to a number of different types of traumatic event which may have affected individuals living in the population.

If a disastrous event affects a whole community at a particular point in time (e.g. in the case of a major flood, earthquake, or landslide, or a terrorist attack) then one might expect to see an increase in the prevalence of mental health problems in the population affected. Many studies have been carried out in recent years which have tried to ascertain the impact of disasters on the mental health of survivors, some of which are reviewed below. These can be useful in estimating the burden of psychological distress and mental illnesses which may need to be addressed in efforts to help the survivors of disasters. However, if the aim of such studies is to

Box 3.1   Examples of questionnaire instruments which are widely used to assess the prevalence of PTSD in populations surveys

The *Impact of Event Scale* and the *PCL Checklist* illustrate questionnaire methods to rapidly assess likely levels of PTSD in a population.

Examples of the items from *The Impact of Event Scale* (Weiss and Marmar, 1997), are shown below (*and in parentheses the type of symptom that the item relates to*).

6) I thought about it when I didn't mean to (*intrusive memory*)
7) I felt as if it hadn't happened or wasn't real (*avoidance*)
10) I was jumpy and easily startled (*hyperarrousal*)

Source: Devilly, G. (2005) The Impact of Event Scale (Weiss and Marmar, 1997): http://www.swin.edu.au/victims/resources/assessment/ptsd/ies-r.html [accessed 1.12.07)

The *PCL Checklist*, produced by Weathers and colleagues at the National Centre for PTSD has versions designed for different types of stressful experience (specific events, past events, military experiences). An example of one of the items used to test for intrusive memories of past events is this:

In the past month, how much have you been bothered by:
PCL-C: "Repeated, disturbing memories, thoughts or images of a stressful experience from the past?"

1 = Not at all
2 = A little bit
3 = Moderately
4 = Quite a bit
5 = Extremely

Source: Weathers (2003) PTSD Checklist (PCL, civilian: PCL-C, military: PCL-M, specific: PCL-S), National Center for PTSD http://www.ncptsd.va.gov/ncmain/ncdocs/assmnts/ptsd_checklist_pcl.html?printable-template=assessment accessed on 20.04.09

A number of other questionnaire instruments for measuring the presence/severity of PTSD are discussed, for example, by: Norris and Hamblen (2004); Kienzler (2008: 220–21).

establish whether the disastrous event has *caused* mental illness and psychological problems, then research needs to establish whether there have been *changes* in levels of mental illness, such as PTSD, associated with disasters.

Collecting rigorous information on mental health in the aftermath of a disaster is very challenging (Galea et al., 2008). Usually the location and timing of disasters cannot be predicted precisely, so it is not feasible to plan survey work in advance, and in the chaotic conditions that often follow such events there is limited scope for conducting research (indeed it may seem to be a low priority compared with responding to more urgent health issues). On the other hand, serious mental health effects of disasters may be felt some time after the event, so it can be useful to follow up populations a while after the initial disruption of the event. In order to assess change in mental health, one would ideally need to have information about mental health in the affected population *before*, as well as after the disaster struck, but this is rarely possible, due to the unpredictability of the event, unless information has been collected previously for a different reason, or routinely collected data sources can be exploited, relating to the period before and after the disaster. The latter strategy was employed, for example, by Hansen et al. (2008) who examined hospital admissions for older people in heatwaves, as compared with non-heat-wave periods. They showed that hospital admissions increased by 7.3 per cent during heat waves for illnesses including symptomatic mental disorders, dementia and senility, mood disorders, as well as neurotic and stress-related conditions). Another strategy may be to compare the post-disaster health status of the population from the affected area with similar populations in unaffected areas. However, one of the sequelae of disasters is often dispersal of refugees from the affected area, making it more difficult to track their whereabouts later, so that samples may be incomplete and biased in the way they represent the populations of areas affected by disasters.

Furthermore, the risk of PTSD is acknowledged to depend on individual vulnerabilities and severity of trauma, so that in addition to information on mental health, other characteristics of the population need to be recorded. For example, variables likely to be associated with the mental health impact of disasters such as floods are: age group; gender; social support; wealth and material resources; previous experience of similar events; degree of exposure to the disaster, the amount of loss or damage. Bourque et al. (2006: 143–147) comment that most research in the field suggests that psychological trauma associated with disasters varies in relation to severity of exposure and existence of previous mental health problems (prior psychological fragility may make survivors more vulnerable to post-disaster distress).

Studies also invoke the idea of *conservation of resources* (Hobfoll et al., 1990; Freedy et al., 1992; Smith and Freedy, 2000*)*, focused on the psychological impacts of material or social loss and of psychological losses associated with disruption of reassuring routines, the sense of loss of control of one's life and ability to accomplish one's goals and emotional deprivation of time with loved ones. For example, this model was found in a study by Goto et al. (2006) to be relevant to

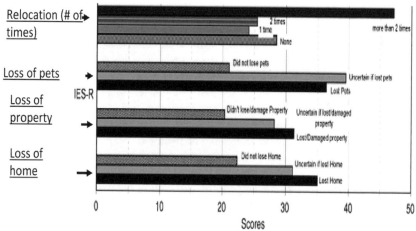

Mean scores of psychological Symptoms: Losses and Relocation

Figure 3.1   Miyake Island disaster (a volcanic event): Relationship between loss and relocation and psychological symptoms of PTSD

Note: Black bars indicate respondents with greatest losses.

Source: Goto and Wilson et al. (2006). Reprinted with permission of John Wiley & Sons Ltd.

psychological impact experienced by a sample from the population of Miyake Island, Japan, displaced in the aftermath of a volcanic eruption (shown in Figure 3.1 from their study, which illustrates the generally positive relationship between greatest loss and relocation and psychological symptoms of PTSD).

In addition, models relating to traumatic events have included ideas about varying individual 'coping styles' (which may be 'positive', concentrating on ways to mitigate distress, or 'negative', involving anger or blame). (For example, see studies of children's coping strategies by Green et al., 1991; La Greca et al., 1996). Also, as discussed in more detail in Chapter 4, research draws on ideas relating to the role of *social support*, which may help individuals to cope with traumatic events, including Norris and Kaniasty (1996) who proposed the 'deterioration deterrence' model, concerned with the extent to which individuals are protected from loss of valued social support mechanisms during times of stress.

Thus, in summary, the most powerful studies to measure the psychological effects of natural disasters at the population level will:

•   have a large enough sample to produce reliable estimates of health status;
•   use well recognised and tested measures of mental health or well designed qualitative techniques to assess mental health of the respondents;
•   assess what people's state of mental health was *before* the natural disaster took

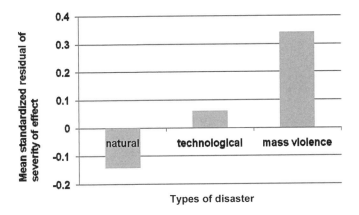

Figure 3.2   Severity of psychological impacts of different types of disaster

Source: Adapted from Norris (2005), Figure 5. Reprinted with permission.

place, as well as *after* (or have a design allowing comparison of those involved in the disaster with a comparable 'control' group who were not affected);

- assess the severity of exposure;
- consider whether other individual or community characteristics might affect the findings.

We should interpret cautiously a large proportion of research in this field because it often fails to sufficiently take account of all of the above considerations (Kessler and Wittchen, 2008). For example, in a recent review of research on the mental health effects of severe flooding, Curtis and Nadiruzzama (2009) found that, while most papers used recognised measures of mental illness, a much smaller number included information on mental health of people in the sample before the flood events, or were able to make comparisons between populations in flooded and non-flooded areas that were otherwise similar.

In the following discussion some selected examples are assessed in terms of what they can tell us about the mental health problems associated with 'natural' disasters, industrial contamination incidents and terrorist attacks.

Mental health effects of 'natural disasters'

Especially since the early 1990s, a growing literature has examined the mental health impact of natural disasters such as storms and floods, severe drought, earthquakes and landslides and volcanic eruptions. Although some reviews (e.g. Norris, 2005) suggest that 'natural' disasters are associated with lower prevalence of mental health problems than disasters due to human activity (see Figure 3.2) there is, nevertheless considerable concern about these mental health impacts. Also, it is important to note that such events have particularly dramatic impacts on populations in low income countries, where there are fewer resources to mitigate or prevent such events than in wealthier countries.

Water as a hazard to mental health: impacts of floods, storms and inundations

Particular attention is paid here to water as a mental health hazard, since this stands in contrast to the depiction in Chapter 2 of water as a benign and 'therapeutic' landscape element. Natural disasters involving flooding and inundation illustrate how water can have a much more sinister aspect. A significant number of studies have examined flooding in river valleys or coastal settings, associated with heavy rains and storms, events such as tsunami, or structural failures such as collapse of dams or defences along coasts and river banks (sometimes provoked by earthquakes rather than meteorological conditions). These are reviewed, for example, by Hajat, Ebi, et al. (2003); Ahern, Kovats et al. (2005, 2006), Galea et al. (2005) Bourque et al. (2006) and Curtis and Nadiruzzama (2009).

Examples of flood-prone areas which have been the focus of many studies include the Mississippi River basin in the USA. Significant flood episodes include those which occurred in mid west USA in spring and summer of 1993 when extensive areas of the Mississippi River basin were inundated. There were multiple flood surges and the floods persisted for around 150 days in some areas. The financial cost of damage sustained was estimated at $20bn. More recently the aftermath of Hurricane Katrina, in 2004, which famously demolished the southern coastal sea defences around the city of New Orleans also caused great consternation because of the widespread and long lasting impacts on the populations of the flooded regions, and gave rise to a flurry of publications about the impacts on their mental health. A different geographical focus of attention has been the regions on the shores of the Indian Ocean impacted by the 2004 tsunami which resulted from an earthquake in the ocean bed. Extensive damage to the coastal region and significant loss of life from the tsunami, as well as the international news coverage of this event prompted a number of studies of the effects of the trauma of this event, both among residents in the area and for visitors from other countries who were tourists in the area at the time the disaster took place.

One of the early studies concerning the mental health effects of floods which has a relatively rigorous design is by Phifer et al. 1990. This study was able to compare mental health for a sample of people over 55 years of age, living in Kentucky, USA, before and after a severe flooding event. Older people who had a prior history of mental health problems were much more likely to report an increase in depression and anxiety after the flood. The statistical models showed that, controlling for other factors, depressive symptoms increased with increasing

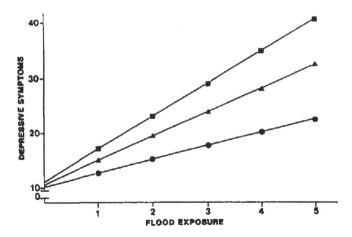

Figure 3.3   The association between flood exposure and depressive symptoms, Kentucky

Source: USA Phifer (1990).

severity of exposure to the flood, and this association was most pronounced among those in lower income groups (Figure 3.3), which supports other findings suggesting that poorer people are most vulnerable to the impacts of natural disaster, probably because they have less financial and material resources to help cope with the effects. Although they might experience material losses which are of lower monetary value than for richer people, they are also less likely to be able to protect themselves through strategies such as insurance cover.

A study of the 1993 floods in mid-west America by Ginexi et al. (2000) also benefits from a longitudinal design, with measurement before and after the flood event. This was possible because, fortuitously, the researchers had studied depression using CES-D in survey of 1,572 residents in Iowa in 1992 before the floods, and they were able to conduct a follow up study in 1993, after the floods had taken place. In 1993 the questionnaire included items on degree of flood impact, which showed that 893 respondents had been directly affected, and leaving a 'control group' who had not directly experienced the flood. Controlling for pre-flood mental state and demographic factors, every one unit increase in the flood impact scale was associated with a typical increase of 10 per cent in the 'odds' (probability) of reporting depression in 1993, 60–90 days after the flood. Ginexi et al., 2000 were also able to analyse the other factors that made people more or less vulnerable. Their study showed that after the midwest flood, people were at *lower* risk of depression if they were: older; wealthier; married. Informants were at *higher* risk of depression if: they were separated from their spouse, they had had depression before the flood; or they experienced greater flood impact.

Kessler, Galea, et al. (2006) were able to use data from *The National Comorbidity Survey-Replication Study* collected in the period 2001–2003 and follow up 1,043 people after Hurricane Katrina. After the hurricane the sample showed significantly worse mental health than before on most measures, but levels of suicide ideation were lower, which the authors interpreted as evidence that there were also effects of 'Post-Traumatic Growth'.

The hurricane related floods of 2004 in the USA were followed up in numerous other studies, but these were less rigorous than the research described above, since most were only able to examine mental health after the hurricane, some involved quite small samples, and being, in many cases, opportunistic studies of particular groups of people receiving medical and welfare support, they were often rather specific in terms of the population groups they represented (Curtis and Nadiruzzama 2009). Considered together, however, they are interesting, for the way that they explored different mental health outcomes, compared different groups of people whose experiences of the disaster varied and explored the relationships reported with other factors which seemed to increase or reduce 'vulnerability' to the risk of mental distress and illness.

Several studies investigated experiences of specific socio-demographic groups. One example is reported by Acierno, Ruggiero, et al. (2006) who conducted a random-digit dialling telephone survey of 1,130 older adults (over 60 years) and 413 younger adults following Hurricane Katrina in the USA. Compared with younger people, older people reported fewer symptom counts of PTSD, major depressive disorder and generalised anxiety disorder. Variation in factors such as age, sex, social support, displacement, financial losses incurred, perceived positive outcomes, and self-rated health status explained much of the variation in MH outcomes, but the associations were different for older than for younger people in the sample. Economic consequences were especially important for older adults. In an ethnically specific study, Chen et al. (2007) collected data on 113 Vietnamese-American hurricane survivors over a 12-month period starting six months after the 2004 hurricane Katrina event, and measured outcomes such as PTSD (the Impact of Events Scale), quality of life (using the SF36 scale discussed in Chapter 1) and a measure of perceived social support. Financial difficulty was the strongest predictor of distress, associated with greater PSTD and worse mental health scores. These studies therefore support some of the theoretical models discussed above including, the 'conservation of resources' model and conceptual frameworks based on the importance of social support such as the 'deterioration deterrence' model. The significance of perceived social support for the mental health outcomes of disasters is relevant to the discussion in Chapter 4 of this book concerning the 'buffering effects 'of social support.

Other studies have investigated how major flooding events have effected women and children. For example the mental health impact of the Indian Ocean Tsunami for a sample of 325 adolescents and their mothers is reported by Wickrama and Kaspar (2007) (see Figure 3.4) in two coastal villages in Sri Lanka, one month after the tsunami. Their study used questionnaire measures of PTSD

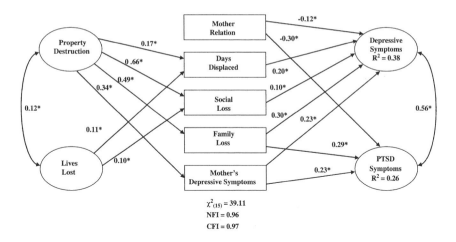

Figure 3.4   Structural equation model showing relationships example of SEM results between aspects of exposure to the Indian Ocean Tsunami and psychological health

Source: Wickrama and Kaspar (2007) Reprinted from *The Lancet* with permission from Elsevier.

and depression, analysed using structural equation modelling. Family losses were associated with both maternal depression and young people's symptoms of PTSD. Also, depression affecting the mother was associated with worse symptoms of PTSD for her child, while children who benefited from positive and supportive maternal relationships had lower levels of post traumatic stress.

Relating more particularly to models concerning individual coping mechanisms, Acierno, Ruggiero, et al. (2006) found evidence of some positive psychological outcomes of exposure to hurricane Katrina, and this relates to the debate about 'Post-Traumatic Growth' (PTG). A study of the post-Katrina experiences of children by Cryder et al. (2006) focused more specifically on PTG in children, looking for evidence of positive psychological change as a result of the struggle with major loss or trauma, which is thought to relate to ruminative thinking, the social support available to them, and their competency beliefs. They interviewed 46 children from schools in the area worst affected by the hurricanes, about 12 months after the hurricanes struck. They found considerable variation in scores on a psychometric scale used to measure PTG. Children showing stronger PTG also had higher scores on measures of their sense of competency. This may mean that some children have more resilient personalities and confidence in their ability to cope with the stress of the disaster, and it is interesting that this sense of competency was also linked to their reported levels of social support.

Box 3.2   Qualitative studies of experience of flooding in England

Studies using in-depth interviews and similar qualitative techniques can illuminate the experience of severe flooding events and help us to understand how these experiences impact on mental health. This can inform conceptual models for more extensive studies which explore the pattern of similar experiences in the wider community.

Tapsell and Tunstall (2007) conducted a study in the English towns of Banbury and Kidlington following serious floods in 1998 using a qualitative methodology involving discussion groups. Carroll et al. (2009) report on another example of psychological impacts experienced after flooding, based on a study in Carlisle, UK. This study also employed in depth discussion group methods.

Some (strikingly similar) comments of participants in two separate studies express clearly the sense of losses which were both materially and emotionally significant, consistent with the 'conservation of resources' model. As well as the immediate trauma and loss, respondents reported strain placed on their personal relationships and their mental health by the protracted process of repairing the damage and reconstituting the home (referred to by Carroll et al. (2009: 4) as 'reconstruction, restoration, return to routine'). These accounts of negotiations with insurance companies and dealing with builders over repairs in the aftermath of the flood, help to explain why the psychological impacts are often felt for a longer period after the disaster event. This suggests that disasters such as floods should be viewed as ongoing stressors, rather than short-run disruptions to daily life.

"That was it – gone, all that you'd worked and saved for...was just going."
"Financially it drains you...because you're needing things all the time – for the children."
"Seeing all my home just covered in rubbish...it's like you feel as if your life has been rubbished."
"...what price husband and wife rowing just because you were flooded five years ago and nobody's done a thing about it."

[Extracts from informants' comments, from Tapsell and Tunstall (2007), pp. 103–4 (abbreviated from their original)]

"I worked for everything that I had in my house..."

"We had...a lot of personal things you can't replace...even our wedding certificate."

"I've lost interest in my house...it's not home anymore."

"When the assessor came and said 'you've got to get your electrical items from...(an internet company)'...and it was like 'I don't want them' (goods) from them."

[Extracts from informants' comments, from Carroll et al. (2009: 3–5) (abbreviated from their original)]

Tapsell and Tunstall comment that some of the quotes recorded by these researchers suggest manifestations of anxiety, as suggested by psychological models of trauma induced morbidity and the following examples of their material suggest symptoms of hypervigilance or avoidance that are considered typical of PTSD. Again, very comparable comments are reported by Carroll et al. (2009: 5), who draw attention to the close correspondence with PTSD symptoms.

"The anxiety is still very much there...It's just this awful sense of menacing foreboding that keeps happening every time there's heavy rain."

[Quote from a female respondent 18 months after the flood: Tapsell and Tunstall (2007: 105) (abbreviated from their original)]

"...I felt like a robot. I cried but it was all more a dream...like I just didn't have any emotions...I just wanted to forget."

[Quote from a male respondent 5 months after the flood: Tapsell and Tunstall (2007: 106) (abbreviated from their original)]

"...I can still see things floating, that's in your mind and you can still see the water..."

"...can't have these feelings every time it rains...it affects your family because you are too busy pacing about looking at rivers and looking out of the windows..."

[Quotes from respondents in Carlisle, concerning ongoing feelings of anxiety, reported by Carroll et al. (2009: 5) (abbreviated from their original)]

Other research has used qualitative methods to understand the impact of flooding on mental state, and while these studies cannot provide any useful information on the prevalence of mental illness, they do illuminate the ways that survivors perceive their experiences and the ways that these impact on their state of mind. These individual studies are specific to local contexts, and are based on qualitative methods that are not designed to be more widely generalisable, or comparable with case studies elsewhere. However, the parallels in the reported responses noted here suggest that the compilation of combined evidence from such interpretations may be suggestive of rather consistent patterns of experience, supporting and complementing the results of extensive surveys. This is interesting in light of proposals by commentators such as Knoblauch et al. (2005) who argue that the power of qualitative research more generally might be enhanced by more coordinated strategies to combine findings across ethnographic case studies, rather than considering each in isolation. Curtis (2008) suggests that this may be important for the potential of qualitative research to inform public health policy (see Chapter 8 of this book).

The examples shown in Box 3.2, for example, illustrate findings from England that are consistent with statistical survey findings from other countries, considered above. While there seem to be several parallels in the findings of the two studies, suggesting that some aspects of the results might apply to rather different parts of the England, it is not possible to extrapolate with any certainty to other contexts. Some of the same broad kinds of issues may find different expression in other cultural settings. For example, Rashid, and Michaud (2000) report the distress and social discomfort caused to adolescent girls in Bangladesh by conditions in refugee camps where they were living after being displaced from their homes by floods. For the girls in this study distress was caused by difficulties in adhering to the behavioural customs of modesty, closely chaperoned contacts with people from outside the family and rituals for maintenance of purity and cleanliness. This finding may be a culturally specific instance of the more general problem also highlighted in Box 3.2 concerning the impossibility of maintaining normal social relationships when live is disrupted by a natural disaster.

Thus the evidence of studies of water-related 'natural disasters' highlight the potentially harmful aspects of water as a natural element. From a geographical perspective these studies are particularly interesting because they show the interactions between individuals, social structures and environmental conditions. Curtis and Nadiruzzama (2009) argue that research in this field needs to adopt a conceptual model like the one shown in Figure 3.5 in order to appreciate the complexity of these relationships. Reciprocal interactions between the physical and the social environment influence the risk of flooding as well as the social impact of, and response to flood events. Broad scale social factors, such as national flood protection and emergency response systems, relate to local flood hazards and to the social and material resources to address disaster events in specific communities. Individual factors such as socio-demographic characteristics,

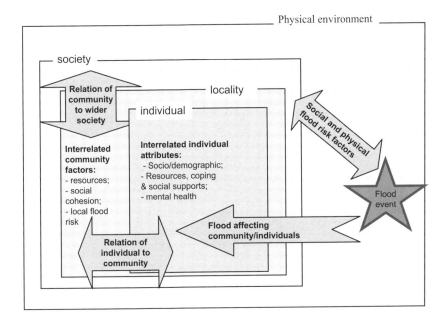

Figure 3.5 **A socio-ecological model of the key relationships influencing the associations between flood events and mental health (after Curtis and Nadiruzzama, 2009)**

economic and social resources, prior mental health and psychological resilience also interact with community factors to influence variations in individuals' mental state within local contexts.

Environmental pollution episodes and mental health

Many environmental disasters are directly caused by human activities and these can also have importance for human mental health. Short run pollution episodes or long term accumulation and persistence of contamination can both have significance. Where contamination incidents occur suddenly, they may be partly associated with PSTD, anxiety and depressive symptoms similar to those discussed above in relation to 'natural disasters'. Longer term ecological problems may also have serious implications for mental health, through more insidious effects of longer term stress, more akin to the problems of the protracted, post-disaster 'clean up' phases following major floods, discussed above. Some selected examples discussed in this section illustrate the international scope of the literature in this field and the kinds of associations between environmental threats and mental health that are typically found. For a more comprehensive review, the reader may wish to consult, for example, Norris et al., 2002a, 2002b) who review the psychological impacts of 54 studies of 'technological' disasters.

Crighton, Elliott et al. (2003) report on the psychosocial health and wellbeing of the populations surrounding the Aral Sea in Karakalpakstan, Uzbekistan, where irrigation systems used for cotton production had resulted in a very severe ecological disaster, including a two-thirds reduction in the volume of water in the Aral Sea and decimation of the fish stock, while also resulting in very high concentrations of pesticides and herbicides in the soil. The physical health impacts of this environmental disaster are very significant, and levels of mental distress also appeared to be relatively high, probably precipitating further physical health problems such as cardiovascular disease due to stress. Based on a conceptual framework illustrated in Figure 3.6 (partly derived from literature by authors such as Folkman et al. (1986) on coping resources – see above), Crighton, Elliott and colleagues hypothesised that the psychosocial impacts of the disaster would relate to the source and nature of contamination, the characteristics of the individual, and also factors in the wider community and society. The research collected information on health which included reported somatic symptoms and stressful life experiences for over 1000 people living around the Aral Sea. The levels of somatic symptoms were found to be relatively high. The authors point out that this was comparable with above average levels reported in an American study of people living close to Three Mile Island, Pennsylvania, USA, at the time of an accident in a nuclear reactor there (reviewed by Norris et al., 2002a). Women were more likely to report symptoms and concern about the environment than were men, which seems consistent with a number of other studies suggesting that there are gender differences in the psychological impacts of disasters (Norris et al., 2002a). The social support networks available to respondents and their satisfaction with their social contacts (especially informal connections with family and friends) was also found to be significant in the Aral Sea example, and this is consistent with theoretical models predicting protective effects of social support such as the 'deterioration deterrence' model discussed above. It is interesting to consider this finding in light of findings from Hamilton Ontario by Wakefield and McMullan (2005), discussed in the following chapter, which showed that residents of a polluted area strongly emphasised social support and social networks in their community. On the other hand, in Karakalpakstan, stronger involvement with more formal community networks was associated with greater levels of concern and stress, perhaps because the most environmentally concerned individuals were most active in their local communities, and this belies some of the arguments about positive mental health effects of social capital discussed in the following chapter of this book. While the researchers had expected those living closest to the old shoreline of Aral Sea to be most affected, this was not in fact the case, as worse psychological health was reported by those living further from the coast. The authors suggest this may have been due to displacement away from the shoreline of communities that had previously relied on fishing in the Sea, and also those living further from the shore would have been more recently affected by

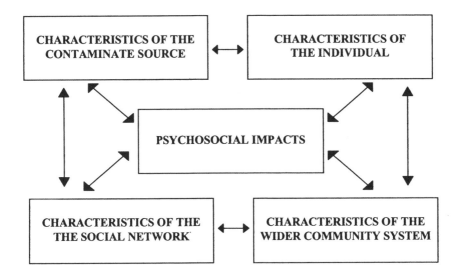

Figure 3.6   Conceptual framework of factors influencing the psychosocial impacts of environmental contaminants

Source: Crighton, Elliott et al. (2003) Figure 2, p. 554. Reprinted from *The Lancet* with permission of Elsevier.

the environmental disaster, so may have had less time to adapt. This underlines the challenges involved in tracking populations involved in disasters over the extended time period during which after-effects are likely to be felt.

Further research on the psychological impacts of environmental disasters in Eastern Europe includes a review by Yevelson et al. (1997) of the mental health impacts of the radiation contamination incident which occurred after an accident at a nuclear power station in Chernobyl in 1985. Summarising reports published in Russian by authors such as: Tarabrina et al. (1994); Napreyenko and Logahanovsky (1995); Rumyantseva et al. (1996), Yevelson and colleagues conclude that among those affected by the Chernobyl disaster, mental health problems included health related anxiety and somatic symptoms in particular, especially for mothers of young children born after the disaster, who were concerned about the impacts of radiation on their babies, and among those most directly exposed to hazards, such as the workers involved in responding to the disaster at the site and those who were evacuated from the area surrounding the nuclear power plant. PTSD was not reported as a significant problem (possibly this was in part because until the mid-1990s Russian psychiatrists worked with different diagnostic criteria than those in the west and so would not all necessarily recognise PTSD). Important issues for mental health after the disaster included the confusing and contradictory information that circulated after the accident from formal and informal sources, which increased stress levels among the exposed populations, and continued to cause concern some years after the disaster happened because of varying and sometimes unreliable accounts of the likely long term health impacts of radiation exposure. These results are interesting to compare with evidence of psychological impacts of the Three Mile Island nuclear accident in the US. Norris et al. (2002a: 222) reviewed relevant research, which suggested that severe psychological impairment was not associated with the accident, but that chronic stress due to residual uncertainty about the risks was significant.

The literature also suggests that people are more disturbed by pollution in their residential area when they feel they disempowered with respect to the cause of, or the response to pollution. Beehler et al. (2008) report evidence for the significance of mastery and controllability of the effects of the disaster for mental health of those in the region influenced by the Chernobyl disaster. A study from Belgium by Vandermoere (2008) seems to illustrate a similar issue with respect to a less spectacular instance of environmental pollution. This study explored the attitudes of residents in a middle class residential area built on a 'brownfield' site, where it was found necessary to decontaminate the soil due to residual industrial pollution. The level of concern among these residents was more strongly associated with the degree to which they felt able to participate in the plans for decontamination than with the level of pollution at their property. One of the causal pathways through which pollution incidents may impact the populations involved may therefore be the undermining of one's sense of autonomy in, and control over one's home environment. The psychological importance of such perceptions of home are discussed further in Chapter 6 of this book.

These studies therefore all seem to illustrate, in different ways, the complex interactions anticipated by Elliott's diagram (Figure 3.6). The nature and severity of exposure to environmental disasters is evidently a factor in the psychological response, and significant environmental disasters seem to generate distress and anxiety which can be quite persistent and significant for health. However, a number of other social and psychological processes intervene to mediate and moderate the disaster impact, making it important to consider socio-geographical contextual variables as well as individual exposures, vulnerabilities and resilience that are important for mental health outcomes.

Conflict, terrorism and violence

Mental health implications of exposure to mass violence, including war and acts of terrorism can also be considered under this broad heading of 'disasters' (see also discussion in Chapter 5 about more individual experience of violence in one's community). Norris (2005) reported from a review of research that studies of people who had experienced mass violence tended to show greater severity of psychological morbidity than victims of natural disasters. As for other types of disaster, the mental health effects varied according to the level of exposure and were more severe among people who had rather direct experience of such events.

Several studies have been conducted among refugees from war zones, especially those who have escaped to other countries and come to the attention of the health and welfare services in the places where they seek asylum. Robertson et al. (2006) studied a sample of over 1000 Somali and Oromo refugee women living in America. Most had experienced losses and reported traumatic experiences involving exposure to violence including torture. Their circumstances were made worse by experiences of displacement and isolation from their original social networks (see also the discussion of these in Chapter 4 of this book). Robertson and colleagues found relatively high levels of PTSD using the PCL checklist (see Box 3.1). They found that symptoms were worse for older women with more responsibilities for family members and with less formal education and ability to speak English.

Working in the UK, Bhui et al. (2006) also conducted survey work with Somali refugees, using rather different techniques. They refer to research suggesting that an exclusive emphasis on PTSD may not be appropriate and they focus on more general aspects of mental health and psychiatric morbidity such as anxiety and depression, as well as PTSD, using the MINI (Mini-International Neuro-Mental Interview), a neuropsychiatric Interview schedule which covers common psychiatric disorders as well as those due to trauma. Their research is an example of 'network sampling' designed to tackle the challenges involved in recruiting samples from rather specific populations such as refugees, who often comprise marginalised, hard to reach minority groups in the general population. Network sampling complements routine sources such as medical registers by recruiting respondents through the social networks and key locations used by the target group. In this study, Somali refugees were contacted partly through community health registers, but also via personal contacts known by the Somali researchers leading the work and via cafes, mosques, and higher education institutions where Somalis were known to congregate. The method is interesting here for the way that it was adapted to the socio-geographical life spaces of the respondents. The research showed that common psychiatric disorders were more prevalent than PTSD symptoms in this group. Those who were asylum seekers and were recruited from primary health care registers had worse health. There also seemed to be evidence that mental health and wellbeing was influenced by the degree to which the refugees were able to establish socio-geographic stability and integrating roles within the host society. The research revealed high levels of mobility within the host country experienced by these refugees after they had arrived in Britain. This was seen by the respondents as detrimental to their health and wellbeing and made it more difficult for services to provide continuity of care (Warfa et al. 2006). Also those who were engaged in employment or education seemed to have better mental health.

These two studies of Somali refugees suggest that the long term impact of the traumas of violent conflict and flight are not only dependent on the nature and severity of these stressful events, but also on a number of social factors and aspects of the geographical context in which refugees find themselves subsequently.

Significant risks to mental health are associated with terrorist attacks. Some recent terrorist attacks seem to have impacted quite extensively, associated with variations in the mental health of people who were only indirectly involved, as well as on those most directly affected. This was demonstrated by research following the terrorist attack on the World Trade Center in New York on September 11, 2001. Chu et al. (2006) demonstrated from a national survey that an emotional response to this event was apparent in the American population generally, with women and certain minority groups expressing greater distress than white males. Galea, Vlahov et al. (2003) also found some evidence of a relatively high level of PTSD symptoms in the population of New York City a month after the attack, especially, but not exclusively, among those living closer to 'ground zero', the World Trade Center site. However, this seemed to dissipate over the following six months.

In addition to those caught up in terrorist attacks, there is evidence that those who are involved in trying to help survivors and reconstruct stricken areas can suffer harm to their mental health. Stellman et al. (2008) report that over 40,000 people were involved in rescue, recovery and clean up operations in the wake of the attack on the World Trade Center. They point out that these workers were subject to distressing experiences involving dealing with human bodies and personal effects of people who had died in the attack, and many were working long hours in arduous and health-threatening physical conditions. The nature of the work as well as the shock occasioned by the event itself therefore seemed likely to produce traumatic impressions on these workers, as well as on people who had been caught up in the attack itself. In a study of 10,000 of these workers, it was found that levels of PTSD among rescue workers were high, and comparable with those among people living in the area close to 'ground zero' surveyed by Galea, Vlahov et al. (2003).

Neria (2008), reviewing these and other studies comment that we have relatively little information on the long term effects on mental health of terrorist events, though it seems clear that after an initial surge of distress the levels of morbidity tend to decline, suggesting a relatively short term effect in many cases (though for some people serious post traumatic effects may be persistent in the longer term). On the other hand it is notable that these short term effects are quite widely felt in the society, not exclusively by those who have directly experienced the event. In Chapters 4 and 8 we review the more general question of how these events, together with more persistent daily threats to security, both at the local level and the wider, global scale influence one's general sense of security and risk in the world.

Placing disaster experiences in context

All of the examples considered above suggest that socio-geographical context matters for the mental health of survivors. Not only does geographical setting critically determine the likelihood and severity of exposure to the immediate impact of the disaster, but also several factors in the subsequent social and material environment influence the risk of PTSD or more common mental disorders which can persist for a long period after the event. Studies like those reviewed above may help decide what measures should be put in place to help protect mental health of people affected by disasters such as floods and tsunamis. The research has informed strategies for flood response to reduce mental health impacts, such as those proposed by the WHO Inter-Agency Standing Committee, which was established 1992 to strengthen coordination of humanitarian assistance, and includes heads of UN and humanitarian organisations. This committee has published 'Guidelines on mental health and psycho-social support in emergency settings' (IASC, 2007), aiming to improve the effectiveness of mental health care in the aftermath of disaster events.

Chronic material poverty and other stressors in the physical environment

The examples considered so far in this chapter show that disasters produce dramatic changes in the material environment, with potentially serious mental health effects, some of which have long lasting effects. However, less spectacular but persistent and long term exposure to material poverty and degraded physical environment may also contribute significantly to the burden of distress and mental ill health in populations around the world. We therefore complete the discussion in this chapter with a consideration of some examples of more common place, day-to-day environmental exposures to material hazards for mental health.

While these may seem less remarkable, because they are part of the routine of daily life for many people, in total they will directly affect more people than are directly impacted by 'disasters', so they have potential to contribute significantly to variation in the overall burden of illnesses and distress in the population (Schell and Denham, 2003). These chronic exposures therefore deserve serious attention. Some of the most basic human requirements are for food and shelter but adequate provision of these essentials are denied to large numbers of people worldwide. Although the physical health effects of such deprivation are perhaps most obvious, there are mental health effects too, as discussed below. Even for those living in otherwise relatively affluent and comfortable physical surroundings, there are questions over whether certain material aspects of environment, such as ambient noise levels or length of daylight hours may be significant for our mental health. Here we consider examples of some of the evidence for mental health effects of these kinds of chronic material stressors.

Prenatal malnutrition, famine and psychiatric disorders

Maternal health and conditions for the foetus in the womb are important for physical and neural development and are known to be significant for child and adult mental health after birth. A number of studies have suggested an association between maternal malnutrition during pregnancy (as well as stress and trauma) and the occurrence of subsequent psychiatric disorders in the child, such as schizophrenia and depression (Bennet, and Gunn, 2006). Some of the research in this field relates to differences in prevalence of schizophrenia among children born during periods of famine, and two famine episodes in particular have been quite extensively researched: the Dutch 'hunger winter of 1944' and the famine which occurred in China in 1959–61, associated with the collapse of agricultural productivity, which was an inadvertent outcome of the 'great leap forward'. The Dutch Hunger Winter, occasioned by severe food shortages in the closing period of the second world war, has been considered in a number of studies which suggest a link with subsequent levels of schizophrenia and major affective disorders such as severe depression and bipolar disorders, and antisocial personality disorders (Hoek, Susser, et al., 1996; Brown et al., 1995, 2000; Neugebauer, Hoek, et al. 1999). The Chinese famine of 1959–61 also shows associations with mental disorders like schizophrenia (e.g. Neugebauer, 2005). Song, Wang, et al. (2009) demonstrate that this association is clearer for those living in urban areas at the time of the famine than for those in rural areas (possibly because the survival rate was very low for those most severely affected by famine conditions in the rural areas).

These famine events are relatively rare occurrences in the countries and the time periods for which they were observed and it is because they were exceptional that these epidemiological studies are possible. They perhaps belong more properly to the previous section of this chapter, relating to 'disaster' events. We should however, consider that severe food shortages are common in other parts of the world which have not been subject to such research, so that the impacts of malnutrition on mental illness may have much wider significance. Some other studies also show that chronic malnutrition in terms of the type, rather than the quantity of food consumed may be important, for example, the importance of Vitamin D for healthy brain function has been considered by Huotari and Herzig (2008) and also by Cherniack, Troen, et al. (2009), who suggest that vitamin D deficiency may be an element in risk of dementia in older people.

Poor housing and quality of residential areas and mental health

A large number of studies have examined the associations between mental health status and housing conditions for people living in buildings and areas which are physically decrepit and unattractive and which may expose residents to physical toxins or other dangers. These have been reviewed, for example, by Evans et al. (2003); Evans and Lepore (2008); Freeman (2008). Evans et al. (2003) reviewed evidence suggesting that mental health was worse on average for those living in multi-dwelling homes, on higher floors of high-rise buildings or in housing of generally poor quality, than for those in single family, detached houses and homes of a better standard. They suggest it is possible that greater isolation in high rise dwellings and lack of access to outdoor space and safe play space for children might contribute to worse psychological health. Other factors that might produce general distress or anxiety were postulated, including: the way that housing reflects one's social standing and sense of self-esteem; worries about structural and other hazards in the home; lack of control over maintenance. If high rise buildings are less attractive architecturally or seen to be more exposed to risk of crime, this might also contribute to mental health differences. However, authors such as Evans et al. (2003) and Thomson et al. (2006, 2008) point out in their reviews that it is difficult to draw firm conclusions about the nature and direction of causal pathways producing these associations given the limitations of the evidence available. Several of the studies reviewed did not control sufficiently for socio-economic and housing tenure differences associated with type and quality of housing, many of the studies were cross sectional in time, so one cannot determine whether poor housing may have preceded poor health, and several of the studies relied on self reports of both mental health and housing quality, which is problematic because people who are depressed or anxious might be more likely to view their housing negatively.

More rigorous studies of mental health and housing, including studies using a longitudinal approach to examine whether mental health changes after housing improves, and controlling for at least some of the other socio-economic variables that might account for mental health differences, seem to give varying results (Thomson et al. 2006, 2008). For example, Evans, Wells, Chan and Saltzman (2000) report a study in which housing quality was assessed by trained researchers, rather than residents reports, and were able to measure mental health before and after people had moved to housing of a different quality. This study showed that, controlling for income poverty, mental health measured on the Psychiatric Epidemiology Research Instrument was worse for those in housing rated as of poorer quality in terms of factors such as structural quality, indoor climate, cleanliness, privacy. Also those who moved to better housing subsequently were found to have improved health.

In contrast, a study in England by Thomas et al. (2005) examined distress measured by the General Health Questionnaire before and after refurbishment and improvement to the quality of housing for their sample, and compared them with another group whose housing had not been improved. There was no evidence of an improvement in mental health following housing refurbishment. The authors suggested that other aspects of the environment, apart from the physical quality of housing, seemed more important for respondent's psychological health. This emphasises that associations between poor housing and health in general may be quite complex and need to be considered in relation to other aspects of living conditions, a point which is also suggested by other research findings. For example, Evans, Hyndman, et al. (2000) found that psychological distress due to worries over other aspects of life, such as work, had links with measures of physical health such as asthma that were as strong as the associations with the physical aspects of housing.

The environment in and around the home may be important for mental health in various other ways. The social environment in a neighbourhood can have significance for mental health, as discussed in the next chapter, but physical environmental contamination of various types can also play a role. One issue is risk of physiological damage to brain function from some forms of environmental pollution (e.g. Schroeder, 2000, has produced a review of risks for cognitive development and mental retardation among children associated with the effects of neuro-toxins such as lead and methylmercury. Exposure risks may interact with individual vulnerabilities associated with genetic makeup, personal and family resources, and social conditions to produce varying risks of mental retardation and cognitive development problems for young children exposed to these neuro-toxins. He argues that the New Morbidity model first proposed by Baumeister and Kupstas in 1987 (e.g. see Baumeister et al., 1991) (Figure 3.7) provides a useful framework to express these interactions between individual and environmental factors.

Lead exposure due to traffic pollution involving lead additives in petrol fumes and, in older housing, lead pipework in the water supply system, has long been noted as a problem because of the neurological damage it causes to children. Recent research in Mexico city (Tellez-Rojo et al. 2006) suggests that even rather low levels, which were previously thought to be safe (e.g. Davidson and Myers, 1997), may carry some risks. Koren and Butler (2006) comment on the implications of economic and industrial growth in lower income countries which implies a major increase in the numbers of people exposed to these pollutants in their residential environments worldwide. Other studies focus on the health risk to infant's neurological health associated with pollution from pesticides which may be problematic in rural areas (e.g. Cohen, 2007). Ambient noise as a factor in mental health and cognitive development of young people has been investigated by Stansfeld, Berglund, et al. (2005), who showed, in a major study of school children in the Netherlands, UK and Spain, that exposure to noise close to major airports was associated with increased sense of annoyance and with reductions in children's cognitive development, even after controlling for socio economic status, education and ethnicity of the children's parents, which might also have been associated with measures of children's development.

Another example of normally prevailing environmental conditions linked to mental health problems is the seasonal variation in sunlight levels, particularly in high latitudes, which is associated with Seasonal Affective Disorder, causing depressive symptoms in some people when daylight hours are very short and there is a lack of bright sunlight (e.g. considered by: Kasof, 2009; Kegel et al., 2009; Steinhausen et al., 2009.)

Thus environmental quality of one's residential setting has significance for mental health, partly for physiological reasons, but also because of the psychosocial aspects and importance of the home space for one's sense of identity and social standing (an issue returned to in Chapter 5). Thus once again, we note that specific risks in the physical environment need to be considered in the light of complex interactions between individuals and their residential setting.

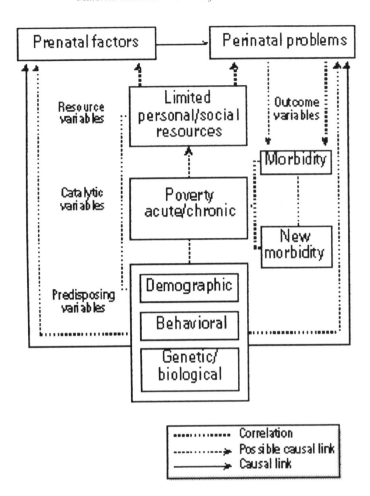

Figure 3.7   The New Morbidity model of risk for mental retardation proximal variables

Source: Schroeder, 2000.

Conclusions

The research reviewed in this chapter shows that mental as well as physical health may be at risk from material conditions in one's environment. This is brought to public attention most forcibly in the event of environmental disasters, whether the main causes of these are 'natural' or primarily due to human activity. However, especially with the increasing urbanisation and industrialisation of societies around the world, we also need to consider the aggregated impact of more chronic risks and stressors which also take a considerable toll on mental health and wellbeing. It is possible to measure these effects at the 'population level' (in non-western as well as western countries). The research reviewed here has also repeatedly shown that there are differences in individual's vulnerability or resilience to material environmental risks for mental health, associated with social characteristics of the person and their setting. Much of the research has been carried out in more developed, wealthy nations, and it is sometimes suggested that people in western cultures are more sensitive to environmental threats, while in other cultures, where populations have more experience of environmental disasters and deprived material living conditions, people have developed a greater 'hardiness'. However, a growing body of research is showing that these environmental risks are also important for mental health in developing countries and that socio-economic and physical environmental change in all parts of the world is bringing these issues to the forefront of public health agendas.

Key learning objectives

This chapter aims to provide insights into:

- psychological responses to disaster events, notably post-traumatic stress disorder;
- the challenges involved in researching the mental health impacts of disasters;
- the relationships between chronic material deprivation or contamination and mental disorders.

Introductory reading

For a general statement of the issues of significance for human health care in the event of disaster this is a useful source:
IASC (2007). Inter-Agency Standing Committee Guidelines on mental health and psycho-social support in emergency settings, retrieved 18.04.09, from http://www.humanitarianinfo.org/iasc/downloaddoc.aspx?docID=3981&type=pdf.

For reviews of mental health impacts of flooding and other disasters:
Few R., and Matthias, F. (ed.) (2006). *Flood Hazards and Health*. London: Earthscan.
Norris, F.H., M.J. Friedman, et al. (2002a). 60,000 disaster victims speak: Part I. An empirical review of the empirical literature, 1981–2001. *Psychiatry – Interpersonal and Biological Processes*, 65(3): 207–39.
Norris, F. (2005). Range, magnitude and duration of the effects of disasters on mental health. Retrieved 14.04.09, from <http://www.redmh.org/research/

general/effects.html March 2005 <http://www.redmh.org/research/general/effects.html%20March%202005>.

For thoughtful assessments of some of the theoretical issues raised in this chapter and interesting case studies:

Milligan, C., and Bingley, A. (2007). Restorative or scary spaces? The impact of woodland on the mental well-being of young adults. *Health & Place*, 13(4): 799–811.

Carroll, B., H. Morbey, et al. (2009). Flooded homes, broken bonds, the meaning of home, psychological processes and their impact on psychological health in a disaster. *Health & Place*, 15(2): 540–47.

Tunstall, S. et al. (2006). The health effects of flooding: social research results from England and Wales. *Journal of Water and Health*, 4(3): 365–80.

Chapter 4

Resilience, Social Capital and Social Integration

Summary

This chapter reviews recent developments in the literature on social relations in different settings and their importance for mental health. The focus is primarily on those aspects of neighbourhood social relations which seem beneficial to mental health, with particular attention to studies of how mental health relates to 'social capital', 'social efficacy', social networks and social support at the community level. This research raises questions about how to conceptualise the 'strengths' of social relations in communities and their association with mental health and resilience. Strategies for operationalising the relevant theories and testing them in qualitative and quantitative research are also discussed. The examples of empirical studies considered here include consideration of research on the significance for health of social ties within local communities and of 'ethnic density' of minority groups. 'Healthy' social relationships may help to build resilience to environmental challenges, as discussed in other parts of this book.

Introduction: social environments and mental health: geographical perpectives

Here and in Chapter 5, the discussion focuses on the significance for mental health of the social environment, thought about in terms of social relationships at the community level. This chapter starts with discussion of some of the issues involved in conceptualising the 'social spaces' within which these social relationships are constituted. The discussion then moves on to consider the diversity of theories about 'social capital', concerned with aspects of social relationships which may be considered beneficial for wellbeing at the individual and collective levels. Social capital is interpreted through conceptual frameworks involving ideas about social resources, social efficacy, social networks and social support. These are often argued to be supportive for psychological health of individuals and for population level mental public health. Research in this field raises interesting questions from a methodological perspective about how we can delineate and describe the social environment using geographical information, produced using both qualitative and statistical approaches. This is a contested field and there is a debate as to how far the advantages conferred through certain types of social capital are universal, benefitting the whole community, or selective, contributing to social inequalities. In this chapter we consider examples of empirical research on the significance of community level social relationships for mental health and wellbeing. Some of these focus on specific population groups, and this chapter takes as an example research on the mental health implications of spatial congregation and 'density' of minority ethnic groups.

Conceptualising 'social spaces'

Much of the literature in psychology interprets 'environmental' factors and 'social relationships' in terms of family or household conditions that interact with the characteristics of the individual person to influence their mental health. It is therefore important to emphasise that a geographical perspective also leads us to consider social relationships in the wider community, especially at the local, neighbourhood level, but also in society more widely.

Furthermore, while the focus in this chapter is primarily on processes in small areas approximating to local 'neighbourhoods', these are not the only 'social spaces' of concern to health geographers; from a geographical perspective we can view the 'social environment' as comprising a complex network of social spaces linking individuals at various interlocking scales. We may also interpret 'micro-level' spaces at the scale of the human body in terms of performance of social interactions, and social relationships may also be constituted in much more extensive or physically fragmented spaces at the level of nations, international networks or 'global cultures'.

Social spaces are often discussed in terms of the ideas of 'community' which have a long and very mixed pedigree. Tönnies' (1887) made a distinction between 'Gemeinshaft', a concept of traditional, preindustrial small scale community held together by links of kinship and friendship and 'Gesellschaft', the idea of modern, industrial associations based on legal, contractual ties. A similar distinction can be seen reflected in some of the ideas of social capital reviewed below. Notions of urban communities as 'social areas' are evident in early 20th century work of the Chicago school including writing by Park, Burgess and McKenzie (1925), and a much cited collection of 'community studies' was edited by Bell and Newby (1971). There is no single definition of 'community'. Shore (1994: 98) suggests this is "[o]ne of the most vague and elusive concepts in social science". It is interesting for the discussion below that some recent views on community (e.g. Popke, 2003; Welch and Panelli, 2007) draw on theories by authors such as Nancy (2000) who have critiqued the notion of 'community-being', in the sense of being *part of* a monolithic, enduring social group, and instead emphasise the idea of a 'plurality' of singular individuals who have a shared sense of 'being *with*' each other, which may be rather loose, variable and ephemeral, and is not restricted to a particular geographical locality.

Furthermore, relational geographical perspectives suggest that human relationships and interactions at different geographical 'scales' should not be treated as rigidly hierarchical and independent of each other. Quite often processes operating at different scales interact together in their associations with mental health. For example, as discussed below, some 'neighbourhood' processes relate to individual's mental health indirectly, through their relationship to processes within families, so some reference is made here to social relationships in family settings.

It is often argued that social relationships between individuals are constrained and constituted by 'larger scale' structures operating across whole societies. However, there are processes operating from the 'bottom up', as well as from the 'top down' that are important for the way that human interactions develop. For example, Nielsen and Simonsen (2003) discuss how structures and processes at the large scale may be changed by micro-level social interactions, as well as imposing structural constraints on micro level interactions. In the field of mental health care policy, an example would be the development of national guidelines on how to care for people with mental illness, and how to manage risks that individuals with severe mental disorders may harm themselves or others. These guidelines affect treatment for many people with mental disorders, and as we shall see in Chapter 7, they have significant effects on the settings in which mental health care is provided. However, guidelines on good practice in this field frequently seem to be brought into effect in response to strong public reaction to specific, local, high profile incidents involving individual patients (Moon, 2000).

When considering the 'social environment' we also need to consider that it does not consist exclusively of people and their behaviours towards each other. The social environment is partly 'composed' of groups of individuals with various characteristics, so that from an actor network theory perspective (e.g. interpreted from a geographical perspective by Murdoch, 1997, 1998) individual people are important 'actors' in their own right, but their interactions with each other and the nature and function of other elements in the spaces surrounding and connecting them are also essential to consider. Thus it is impossible to fully understand social relationships operating in human communities without knowledge of the other agents in the local context. For example, the routes and transport systems that make communication more or less possible, the built environment of public and private spaces in which people meet, the institutional structures which frame their interactions, the economic infrastructures for production and consumption that affect social relationships. Context therefore matters for social relationships and the interactions between individual people show variability from one setting to another, generating geographies of social relations. The significance of these for the sustenance of mental wellbeing is the main focus in this chapter.

Conceptualising social relationships: 'social capital' and mental wellbeing

Certain aspects of social environments are thought to contribute to maintenance or recovery of 'healthy' psychological states and wellbeing. These are viewed here through the lenses of a cluster of theoretical frameworks concerning the roles of 'social capital' and 'social resources' in community social relationships. A useful recent overview of these is provided by Fulkerson and Thompson (2008), who present a meta-analysis of the various formulations of ideas about 'social capital'. They comment on the explosion of publications using this concept in the last 10 years, and they highlight two different interpretations of the nature of 'social capital' dominating this literature referred to as 'normative' social capital and social capital as 'resource'. Portes (2000) is among other authors who make a similar distinction.

Authors espousing the notion of what Fulkerson and Thompson term 'normative' social capital, are working in the tradition of arguments first put forward by Durkheim (1858–1917). Durkheim was mainly concerned with what he saw as problematic trends in modern society (e.g. Durkheim, 1993; discussed in Chapter 5 of this book), but he contrasted these with a rather idealised vision of traditional societies, emphasising the value of traditional family relationships and the role of wider society as a sort of 'moral framework', giving people a reassuring sense of a clear and secure position in society. A similar 'moral structure' perspective on society is reflected in what Fulkerson and Thompson (2008; 540) describe as the 'normative' view of social capital, seen as a 'set of features in a social structure that lead to collective action in order to bring about mutual benefit for some aggregate of people'. These features form a pattern of 'normative' social relationships understood in terms of attributes of social organisation such as *trust* and *reciprocity*, *cohesion* and *solidarity*, which are seen as rather universally beneficial to societies. This view was put forward first by Hanifan (1916) and later by authors such as Coleman (1988) and Putnam (2000).

This perspective tends to study social capital by focusing on social organisation, looking for evidence of these beneficial structures and their relationship to various outcomes in society, including psychological health. For example, Putnam (2000) distinguishes between different types of social ties that are important for social capital, which he defines as 'bridging' and 'bonding' ties. Bridging ties provide social connections between different groups of people and they provide channels through which individuals may have access to a wide range of opportunities and possibilities in their social life, as well as sources of useful information to help them achieve their goals. Bridging ties, for example, may help people with access to living conditions which are important for mental health such as education, housing and jobs, and may facilitate access to information and services that are important for health and health care. Bonding ties are social relationships which 'glue' together people within specific social groups, building a sense of solidarity and of belonging, being embedded in one's social environment.

Normative social capital is often represented as a 'bulwark' against challenges and risks for human wellbeing and implies a critique of modern society, invoking advantages of more traditional social structures (in this respect the argument refers to a well established discourse dating back to Tönnies' writing). The title of Putnam's (2000) book *Bowling Alone* invoked the idea of an increasing level of social isolation in American society during the 20th century with individuals participating in many daily activities alone, rather than in groups (even in events such as trips to the bowling alley, which might be considered good opportunities for social gatherings). Accounts such as this one seem to have captured the public, as well as the academic imagination. Perhaps this is partly because they 'touch a nerve' for members of contemporary 'western' societies, such as North Americans and British people, who find that in spite of increasing material wealth, they feel increasingly alienated from each other, lacking a sense of belonging to a social group. (See the discussion in Chapter 5 concerning social fragmentation and *anomie*). Normative accounts of social capital hold out the possibility of an 'alternative' form of 'value' and 'worthiness' in the social realm, comprising membership of social support networks, trust in others and socially reinforced norms of behaviour. Also, 'social capital', is a term which resonates with ideas of 'circuits of economic capital' and economic growth in capitalist political economies, while also highlighting the need to conceptually differentiate 'social' and 'economic' dimensions of social relationships and resources, These normative accounts of social capital suggest it has potential to compensate or 'buffer' those in disadvantaged groups for their lack of economic capital. It can provide a source of resilience against poverty and other stressors.

The second broad perspective identified by Fulkerson and Thompson emphasises social capital as a *resource*. This view is aligned more closely with interactionist schools of thought and perspectives concerned with uneven power relationships and struggle and conflict among social groups. Authors associated with this perspective include Bourdieu (1984) and Portes (1998). They interpret social capital in terms of investments that individuals make in networks of social relationships which provide access to resources that are important for one's social position. Fulkerson and Thompson suggest that some authors adopting this 'resource' perspective (apparently including geographers such as Gatrell and Popay (2004), as discussed below) are uncomfortable with the 'normative' view of social capital, partly because it might be interpreted in terms of ideas about 'cultures of poverty' whereby deprived social groups might be represented as responsible for their disadvantaged position, being culturally averse to helping themselves by developing normative attributes of social capital. They therefore prefer to avoid using the term. Others choose to work with a view of social capital as a 'resource' and concentrate on the ways that social relationships reinforce unequal access to resources and social power for different social groups. Their work shows how social and economic capital may be substituted for each other. 'Resource' social capital often boosts economic position and material living standards for those who are already relatively advantaged in society (social as well as economic factors influence access to education and employment, which in turn benefits subsequent economic and social position).

Theorising the links between social relations and psychological health

While the theoretical literature on 'social capital' does not, therefore represent a consensus, the various discourses suggest diverse ways that social relationships may influence psychological and physical health.

Mechanisms by which 'normative' social capital is expected to improve health, according to Kawachi et al. (1999) and Putnam (2000) include: reinforcement of social 'norms' which encourage positive health behaviour (such as healthy physical exercise and healthy eating) and discouragement of health damaging behaviour (such as excessive alcohol consumption or use of illicit drugs); enhanced access to facilities and activities which benefit health; provision of emotional and practical help to people who are frail and impaired due to poor health; reduction in anxiety and fear, due to improved levels of trust of other people and collective strategies for controlling crime and incivilities; and enhancement of senses of self efficacy and self esteem through being able to play 'worthy' social roles in constructive help giving and other social activities.

Other authors focusing on accounts of social capital which emphasise the distribution of 'resources' tend to emphasise the role of social relations in generating health advantages which are unequally distributed. Examples include Stephens (2008), Carpiano (2006), Bernard et al. (2007). These accounts include discussion of the socio-economic processes giving variable access to resources. Bernard et al. (2007) (see Figure 4.1) discuss 'domains' of social relationships within which different sets of 'rules' determine the distribution of resources. In formal social institutions access is regulated by 'rights' in less formal domains of community organisations and neighbourly sociability, processes of informal reciprocity predominate, and within the economic sphere price mechanisms govern access. Based on a review of the literature Carpiano (2006) argues that health behaviours and risk factors as well as health outcomes are influenced by social cohesion (social connectedness and shared social values) as well as aspects of social capital (social support, participation in neighbourhood organisations, social control and social leverage). He incorporates ideas of 'normative' social capital in his conceptual model while noting critiques of the 'normative' model, and arguing that certain aspects of local social relations benefit some groups more than others. Stephens (2008: 1176) comments that a perspective drawing on Bourdieu (1984)

> ...shifts our attention to the role of the wealthy in perpetuating inequalities...
> The understanding that possession of resources is about unequal social relations
> between groups, and about exclusion of others from beneficial resources, maps
> well onto current observations of health inequalities.

A conceptual model theorising the positive contribution of social environments which is more specific to psychological health is provided by Berkman, Glass and colleagues (Berkman et al., 2000; Kawachi and Berkman, 2001). Their conceptual framework draws on earlier literature, including authors such as Durkheim (see

Neighbourhood

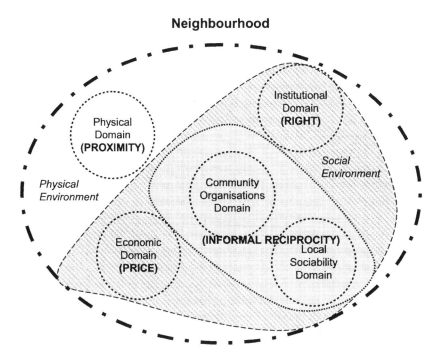

Figure 4.1   Neighbourhood environments and rules of access

Source: Bernard et al. (2007: 1843). Reprinted from *The Lancet* with permission of Elsevier.

Chapter 5 of this book) and Bowlby (1969). It theorises the causal pathways linking stronger levels of 'social integration' with better psychological and physical health outcomes (see Figure 4.2). Social integration is conceptualised in their model as contingent on the socio-economic, political and cultural context of a society. Crucial for social integration are the structure of social networks and the nature of the 'social ties' operating within these networks. These factors in turn influence psychosocial processes that are seen to be important for good health such as social support, influence, engagement and contacts, as well as access to the material 'means to health' such as good housing, healthy employment and adequate income and institutional contacts. These contribute to various forms of social support and positive social influences, contact and engagement with other people and (through social networking which provides a source of information and help) better access to resources such as jobs, housing and facilities. All of these factors are thought to be beneficial for healthy psychological outcomes via pathways such as better self-esteem, self-efficacy and capacity to cope with stress, healthy mood and psychological wellbeing.

It is often claimed that the kinds of social relationships invoked by theories of social capital and social cohesion can be important for mental and physical health

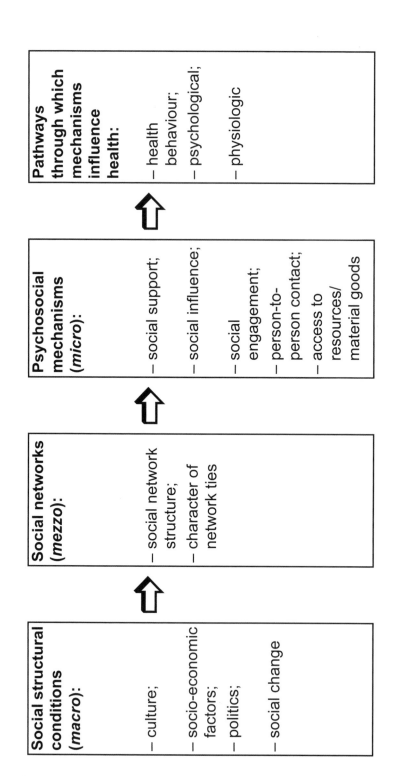

Figure 4.2   Berkman et al.'s model of the links between social integration and health

Source: Adapted from Berkman et al., 2000, Figure 1. Reprinted from *The Lancet* with permission from Elsevier.

because they confer on members of communities enhanced 'resilience' in the face of risks that threaten health. As mentioned in Chapter 1 resilience is considered an important dimension of psychological wellbeing. Research on resilience often draws on writing by Sen (e.g. 1970) who was especially interested in the potential for individuals and social groups to exercise choice and take action to change their lives. His ideas originated from studies of social and economic development in Indian communities with low levels of industrial development and high levels of material deprivation. Although as discussed below, other research emphasises that there are limitations to such agency imposed by social structures and power relations in society, Sen focuses on resilience as a force that can be a valuable source of benefit to wellbeing, even in the face of extreme social disadvantage and material misery. Where social groups organise together to offer each other social support these positive effects can be enhanced for people in the group as a whole.

Representing and describing the social environment from a geographical perspective

In order to collect empirical evidence to test these theories, various research strategies have been developed to collect information about the aspects of local social environment invoked in theories linking social capital and health.

All empirical studies of the social environment face a significant challenge in operationalising the idea of a local community or neighbourhood. As already noted above, while theory suggests that the local context as well as conditions in wider society will be important, it is very difficult to know how to define local 'neighbourhoods'. For pragmatic rather than theoretical reasons, many of the studies considered here have used small areas defined by administrative boundaries, or in some cases by areas defined, by varying criteria, as within 'walking distance' of one's home. As discussed above these may only very roughly approximate functional social neighbourhoods defined by local social connections and practices and will have varying relevance for different individuals living within the same area, according to their individual behaviour and capabilities for movement and communication across space. As observed by McKenzie, Whitley, and Weich (2002) this problem also limits the precision and power of ecological studies of social capital as they relate to mental health.

However socio-geographical areas are defined, a further problem is how to measure the relevant attributes of the social environment, since there are few routine sources of information that can provide suitable indicators. One strategy is to generate original information using intensive research focused on particular communities to explore how the social environment is configured in particular settings. Another way forward is to collect original information in more extensive community surveys which can be used to generate measures of the social environment at the collective level. There are also a few sources of

Box 4.1   An example of an ethnographic approach to the study of social environment and social cohesion using interviews and participant observation

Qualitative research on social networks using in depth interviews and observational techniques is well represented by Cattell (2001, 2004a, 2004b). She recorded perceptions reported by small samples of residents built up by participatory observation, key contacts and 'snowball' samples. The study was located in two areas which were both very deprived urban areas locations in East London. The areas differed in some respects, including aspects of the built environment which in one case was quite traditional and in the other area comprised modern, deck access tower blocks and had a more apparent problem of crime and delinquency. Her approach was 'holistic' (Cattell, 2004b: 145), exploring respondents narratives about their everyday life and experiences. Some of the respondents' comments that Cattell cites are reproduced here, abbreviated from her original account.

Cattell (2004b) interpreted the accounts given, and built up 'typologies' of social networks which typified people's differing experience of the social environment in their neighbourhood depending on their past as well as present connections with their locality and local community. For example, some individuals were found to have very limited networks with small number of other people, classified by Cattell as 'socially excluded' networks. People in this category included respondents who were unemployed, or recently moved to the area, as well as isolated elderly people, single parents and refugees and immigrants moving without their family. Other groups seemed to have especially strong bonding ties with others who had shared similar family or work experiences, comprising dense 'parochial' networks of neighbours or family members, or 'Traditional' networks of tight knit structures connecting people who knew each other through local settings such as schools and workplaces and local clubs. Cattell also identified more extensive and open networks featuring 'bridging ties' more prominently, that were 'pluralistic' (involving memberships of large numbers of groups of different types of people, often not restricted to the local area; observed, for example, in the case of voluntary sector workers), or 'Solidaristic' networks (combining more traditional local links with more extensive connections outside the immediate area).

Cattell's account of interviewees' comments include examples of social solidarity from shared experiences in their working life as well as at home, attachment to place and historical ties extending over generations: "People round here are proud to be working class, they tell you Dad was a docker, and Grandfather was a docker" Cattell (2004: 147). Thus, although communities in the study areas had witnessed the virtual disappearance of local dockyard industries, and they were situated within a very large and diverse labour market in London as a whole, social relationships of those belonging to 'traditional' local social networks could be seen (from the perspectives considered in this book) as 'path dependent'; strongly linked to their lifetime experience of the historical, port based industries which had employed large numbers of local people in the past. In contrast to such accounts of strong social networks, a more socially excluded person said "There is no community spirit at all...There's a lot of mistrust, you worry who you talk to" (Cattell, 2004: p. 148).

Because this kind of research gathers holistic narratives about the role of social networks in people's lives more generally, it is valuable in showing how the social environment is associated with health related experiences and wellbeing. It is important, also, for the way that it privileges the informants' perspectives and draws conclusions from their accounts of the salient features of their social environment, rather than imposing an external view of what is important. Purposive sampling and selection of respondents who share membership of certain social networks allows investigation of particular categories of people who seem to the researcher to demonstrate and explain diversity or similarity among groups within the population, and to explore how connections operate within social networks. Studies such as this one by Cattell are therefore important in showing the causal pathways that can produce and maintain strong social networks, as well as the reasons for diversity of social networks within local areas.

On the other hand, because the samples are small (under 40 people from each area in this instance) and are not designed to be 'statistically representative' of the whole local population, it is important not to assume that these individual experiences are typical of everyone in the local community. Indeed this could lead to unhelpful stereotyping of particular communities, which overlooks the diversity within them. Qualitative studies of this kind do, however, often suggest patterns that may apply in the population, generating hypotheses which may be tested in the wider population using extensive research techniques and statistically representative sampling.

routine information which are considered suitable as proxy measures for some of the dimensions of social capital summarised above.

The intensive, ethnographic approach provides valuable information about the various aspects of the social environment that are important to people and how these are associated to health. The findings from qualitative research cannot be generalised more widely to surrounding places, except in the theoretical sense that understanding relationships in one place may help to frame approaches to similar research in other settings. Some intensive studies do, however, compare different geographical case studies, providing further insights into the local specificity of the relationships involved. For example, Box 4.1 summarises a qualitative approach used by Cattell (2001; 2004a and 2004b) involving interviews and participant observation to investigate social networks in East London. Dines (Cattell, Dines et al., 2008) also used this method to determine the public spaces that had social significance for people living in the area and were seen to be important both for maintenance of 'healthy' social relationships and for their psychosocial wellbeing. Work of this type extends and develops a long tradition of research on the nature of community relationships in East London, extending back to work by Young and Willmott (1957). When repeated intensive studies have been conducted in an area over a long period they have potential to lend strength to each other by revealing how social relationships in local communities, and associations with health and wellbeing have been changing over time, and how these changes relate to wider trends in the socio-political and economic system. Cattell (2004a, 2004b) reports accounts of traditional working class solidarity perpetuated amongst aging dock workers in parts of London, even after their original places of work had closed down. Box 4.1 illustrates the 'path dependence' of local social relationships, rooted in historical, as well as present day experience.

This type of intensive case study considered in Box 4.1 demonstrates considerable variety of experience of social relationships by different individuals and groups living within an area. The diversity is important because it shows us the importance of the interaction between the individual and their social environment. Social experiences are also varied for individuals as they pass through the 'daily action spaces' where their everyday lives are played out.

This spatio-temporal variability and contingency is increasingly being brought to our attention through research which uses 'go-along', walking interviews. This approach involves the researcher accompanying the informant while they go about their daily activities, moving with them through the fabric of their physical and social environment and talking to them about the events and experiences encountered along the way. Various technological devices may be used to support the 'go-along' interview; voice recording to take notes is common and mini camcorders are sometimes used to record sound and vision images of the environment as the researcher and interviewee travel together through the space created by the 'go-along' interview.

Kusenback (2009) discusses the fit between this approach and phenomenological understandings of existence understood through the human experience of moving through places, as put forward by authors such as Merleau-Ponty, reviewed in Chapter 1 of this book. She describes the method as a 'hybrid' between 'sit-down' interviews and participant observation (which often involves 'hanging out' in particular places, rather than moving with an individual informant through space). The approach allows the researcher to focus on the interactions between their informant. Kusenbach (2003: 466) argues that this method is well suited to research on themes including 'social architecture' comprising the "complex web of connections between people..their various relationships, groupings and hieararchies...how they situate themselves in the social landscape", and the social realm of human experience; "...spheres of reality that are shaped by varying patterns of interaction". She also argues that while the 'sit-down' interview may be a good way to derive information about strong, close social ties that are especially likely to come up in a conversation, the 'go-along' interview provides more insight into more incidental, peripheral or 'purely functional' relationships that can also combine to have an influence on the person's experience of their social environment. She gives examples of instances of 'friendly recognition' when people greeted each other in the street, and notes that in some go-along interviews she observed this happening even when the individuals being observed do not know each other personally. Thus the social realm was partly shaped by processes of 'stranger inclusion' as well as reinforcement of connection to known members of one's wider network. Carpiano (2009) underlines similar points about the 'go-along' interview as applied in health geography and he particularly emphasises the relevance of this approach within a relational geographical framework (see Chapter 1). From an Actor Network Theory perspective, this kind of approach emphasises the idea of the person as an actor within a network which comprises many other human and non-human actors interacting with each other variably in over space and time in ways that are important for health.

There are clearly advantages for the researcher in accompanying their informant, as they can make their own observations of the setting and events as well as gathering information about the ways that these are interpreted by the informant. However, it is not essential for the researcher to be physically present in order to collect information about individuals' space-time experiences in their social and physical environment. Related techniques such as 'photovoice' methods hand all the responsibility for field observation to the informants themselves, who carry and recording devises such as cameras or sound recorders so that they can make oral and visual 'notes' of what they see to be important. These are brought back to the researcher and the informant explains the significance of what they have recorded. (There is potential as well for informants to carry portable geographical positioning devices to track their movement over time and space so that their movement can be mapped and linked to their photovoice recordings, in order to be able to precisely locate in space the places they choose to record and comment on.) An interesting example of a photovoice study by Ornelas et al. (2009) which focuses on the environmental factors contributing to social capital is summarised in Box 4.2. This method is often argued to have advantages because it empowers individuals by giving them

Box 4.2   Using photovoice to study social capital

Ornelas et al. (2009) report a study using photovoice techniques in North
Carolina, USA to study African American's perceptions of aspects of
their communities including processes relating to social capital. They
recruited small samples of men as 'navigators' who were inducted
into a role as 'lay health advisor' for their community. They were also
trained in photovoice procedures, including ethical issues involved in
photographing people who might be recognizable. The 'navigators took
part in a 'brainstorming' discussion where they decided on questions
which would be used to frame a number of photovoice assignments
relating to health in their areas. These included assignments relating
to 'things that are healthy about my community' and 'what it's like to
be a Black man in Raleigh'. The participants took photographs in their
community that they thought were relevant to these assignments and
chose pictures to consider in a group discussion where they interpreted
the significance of the images. The researchers facilitated this discussion
using the 'SHOWED' framework (Wallerstein and Berstein, 1988) which
involves answering a number of questions to help participants identify:
what they *S*ee in a photograph; what is *H*appening; how it relates to *O*ur
(participants') lives; *W*hy these issues exist; how one may be *E*mpowered
by this understanding; what one can *D*o to address the issue.

One of the images identified by the participants as symbolising social
networks and social capital was of a church. Ornelas et al. (2009) report
that the church as a local institution was seen as part of a significant social
network with potential to influence health of African American men. A
participant was quoted as saying: "...I see this as the foundation of the
African American Community...and I think it would be the major source
of dissemination of information in an effective way." The researchers also
noted that the church was seen as "a community resource that provided
strength and support during times of need" and "afforded African
American men with an opportunity to connect with other men" Ornelas
et al. (2009: 559). However, this community resource also seems to have
been represented as being undermined by trends in contemporary society.
Concerns were expressed about the waning influence of the church in
rural communities and the fact that pastors in local churches were no
longer necessarily local residents, so that their attention was not solely
focused on a particular community where the church was sited.

This example is interesting from a health geography perspective partly because of the significance of the church as a *material* place representing a social institution and a physical space with an important functional role in social networks that are seen to be important for health. The church building also seems to have a *symbolic* significance, representing certain aspects of the social environment that can be beneficial for psychosocial health. More broadly, ideas from health geography are reflected in the importance attributed to the neighbourhood environment by the participants in this study. Ornelas and colleagues situate their findings in the 'ecosocial' model of health which emphasises the significance of the collective social environment for health. They also discuss the potential of this approach to build capacity among local residents to tackle health issues collectively.

freedom to independently select what to record, so they can choose what for them are important aspects of their environment. The potential of action research of this type as a form of health promotion is relevant to the discussion in Chapter 8. This example also draws attention to the significance for psychosocial health of a religious site. A church building provides a physical space with functional significance, and in terms of Actor Network Theory would be seen as an inanimate 'actor' in the social network. It also seems that the image of a church is used symbolically to signify a social institution. The symbolic value of places for psychological health is taken up in more detail later in this book (see Chapter 6).

Turning from qualitative methods to more extensive, statistical approaches allows a more generalisable picture to be constructed of social capital and social networks in local communities and its relationship to mental health. We are particularly interested, from a health geography perspective, in methods that can be used to generate 'ecological' indicators of the social environment in small areas.

One strategy is to use community surveys, based on samples that are geographically clustered by place of residence, to collect information on respondents' perceptions of social networks and social relationships in their local area. These individual responses can be aggregated over all the people sampled within each area to produce an 'average' view of the social environment at the 'ecological level. For example, Stafford et al. (2003) collected data on perceived social capital in a population survey in Britain, and generated aggregated area indicators by averaging across residents in each of the residential areas in which the sample was clustered. They also used cluster analysis to differentiate varying patterns of aggregated response, finding that areas had varying levels of 'structural' social capital (measured in terms of contacts with other people, participation and social involvement in the community) and of 'cognitive' social capital (sense of trust, attachment, availability of help and tolerance of others). They also found that while there was some association with levels of relative deprivation in areas, the

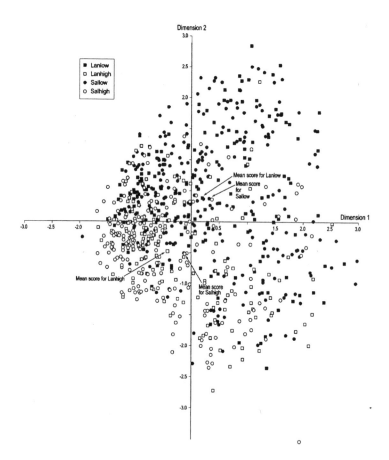

Figure 4.3a   Diagram from multiple correspondence analysis of survey respondents in relation to economic and social capital dimensions

Source: Gattrell, Popay and Thomas (2004: 252). Reprinted from *The Lancet* with permission of Elsevier.

Note: This diagram shows how individual respondents to the survey were positioned in terms of their perception of the social and economic environment in their area. The different symbols distinguish individuals from different local areas and it is clear that not all people from any one area are clustered together. This demonstrates that there is significant variability in individual experience of social capital within communities.

social capital measures seemed partly independent of local economic conditions. In another study of neighbourhood social capital in areas of British Columbia, Canada, Veenstra (2005) used an indicator of social trust prevailing at the neighbourhood level, based on grouped data from a survey of individuals which

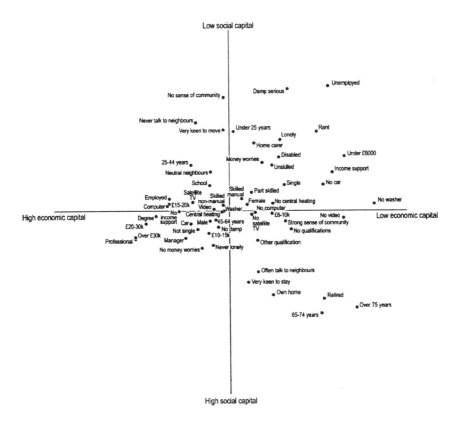

Figure 4.3b   Diagram from multiple correspondence analysis of health determinants distinguishing economic and social capital dimensions

Source: Gattrell, Popay and Thomas (2004: 251). Reprinted from *The Lancet* with permission of Elsevier.

Note: This figure from Gattrell and Popay's multiple correspondence analysis shows how information from various survey items were clustered, reflecting separate dimensions of social and economic capital emerging from the survey responses.

also asked about their health. Western et al. (2005) also used a survey methodology to generate indicators of social capital which were aggregated to produce scores for five Australian communities on four dimensions of social capital comprising formal and informal structures and norms. Gattrel and Popay, Thomas et al. (2003) report a study in Lancaster and Salford, two towns in north west England, where a more deprived and a more affluent area were studied in each town giving four study areas for comparison. The survey recorded respondents' reports of their health and of their local socio-economic environment. This study used multiple correspondence analysis (a form of factor analysis suited to categorical data) to examine the complex interrelationships between different aspects of neighbourhood reported in the survey. The associations in the survey data could be mapped along two 'dimensions' which the authors interpreted in terms of economic and of social capital (see Figures 4.3a; 4.3b). This piece of research is also interesting because it was employed as part of a 'mixed method' approach in which this statistical analysis of extensive survey data was complemented by an intensive qualitative study to explore these relationships in depth, as discussed below (Popay, Bennett et al., 2003; Popay, Thomas et al., 2003). Other approaches to statistically summarising the views of individual people in local communities include 'concept mapping' as applied by O'Campo et al. (2009) to assess perceptions of residents from different areas of the neighbourhood factors significant for mental health and wellbeing.

In studies of relationships between social capital and health, survey data is often aggregated to produce 'ecological' variables for whole communities. However, in analyses which test associations using measures of health and of social capital from the same samples of respondents, measured at the same point of time, it is not really possible to determine whether social capital in the wider environment influences mental health (the problem of 'same source bias'). Subramanian, Lochner et al. (2003) also point out that when aggregating individual data to the area level in order to generate 'ecological' measures of social capital, a more rigorous approach will control for the geographical clustering of other individual attributes which are associated with social capital, including age, sex, and certain socio-economic position variables such as education, which may also be important for health outcomes. They demonstrate an approach using multi-level modeling to produce ecological indicators of social capital differences among areas which are independent of socio-demographic differences.

Another approach which does not require community surveys to generate original data is to collect information from 'routine' sources. These also have the advantage of being recorded independently of any health survey data which might be related to the area social capital measures. For example, in the study referred to above, Veenstra (2005) used two indicators of social capital in neighbourhoods of British Columbia, generated by aggregating to area level data from telephone service directories, adjusted for population size, which measured relative provision of public spaces for activities such as education, culture, sport, religious observances, and relative numbers of voluntary organisations in the area. The limitation of this approach is that there are comparatively few sources of information on relevant aspects of the social environment for comparison across areas, and possible measures are restricted to certain aspects of social capital. For example Cummins et al. (2005) drew up a 'wish list' of proxy measures of different aspects of social capital that they thought might be appropriate to assess the social environment in Britain, in terms of dimensions of social capital. These included: paid newspaper circulation; religious congregation size; religious programme audience share; trade union membership; number of voluntary/charitable sector organisations; volunteering rates; number of blood donations (which are given freely by volunteers to the National Health Service in Britain); financial donations to charitable organisations, voter turnout and political climate. Of these, only the last two were available for all areas of the country in a reliable and usable form. Voting behaviour is often used as an indicator of social capital because the information is readily available from routine public records and it is thought to reflect collective engagement in the social life of the community, but it really only reflects a limited aspect of social capital concerned with civic engagement in formal democratic processes. It does not seem to relate very directly to other activities such as collective engagement in more informal voluntary activities, trust or emotional or practical help available from others in one's community.

From this overview of strategies for assessment of the social environment in terms of levels of social capital, it is clear that, although there has been a rapid development in research to tackle the problem, the methods available for research which aims to compare local communities systematically are still quite limited and hampered by a lack of information. Qualitative research can give us a much richer and nuanced picture of the supportive aspects of the social environment, but one which is restricted to certain places. This needs to be borne in mind as we consider the evidence for associations between the social environment and mental health, since researchers are often working with information on the social environment which is, at best, partial and approximate. Also, because so many different approaches are used it is quite difficult to compare the results of different studies in order to get some overall picture of how local social relations are associated to mental health.

Empirical findings concerning community social relations and mental health

Given all these challenges involved in producing rigorous quantitative information to test associations between community social relations and health at the level of local areas, it is perhaps not surprising that reviews of the empirical research in this field (e.g. McKenzie et al., 2002; Almedom, 2005; DeSilva et al. 2005) suggest that there is little convincing and consistent statistical evidence that community social relationships are clearly related to mental health outcomes. Since these three reviews focus on associations with mental illness rather than more positive aspects of psychological wellbeing, they may not provide a complete picture of the evidence that strong social cohesion and social structures help to maintain good mental health. Some other reviews (e.g. by Morrow (1999) and Ferguson (2006) of evidence relating to children) take a wider perspective on 'wellbeing' or 'quality of life'.

Of these reviews, the most relevant to the present discussion is the relatively recent and comprehensive examination by DeSilva et al. (2005) which concentrates only on studies that controlled for various socio-demographic, and economic factors (such as age, sex, marital status and income or social class), that are likely to be related to the risk of mental illness, in order to explore independent associations with mental illness of aspects of social relationships understood in terms of 'social capital'. They distinguish between, on one hand, studies focused only on associations of individual level measures of social capital and mental health, and on the other hand, a smaller set of seven analyses that met the fairly stringent requirements for the review and also included ecological data such as those discussed above. Of these, six had a multi-level design which combined information on individual mental health with ecological data for the areas where the individuals were living. De Silva and colleagues conclude that these do not show clear statistical relationships between area level social capital and mental health.

A number of further studies were published later than the cut off point for the DeSilva (2005) review. While not all of these meet the same criteria for scientific rigour, they offer further evidence of a rather inconclusive picture concerning the general mental health benefits of social environments where social capital is relatively strong.

Most of the research to date is from Europe (mainly Britain) and North America. However, an increasing number of studies provide evidence from other parts of the world. Findings have now been reported suggesting that trust in other people in one's community is associated with better mental health for socio-cultural groups as different as poor Lebanese women (Khawaja, Abdulrahim et al., 2006) and residents in rural China (Yip et al. 2007). Thus there is some evidence that certain aspects of social capital are linked to mental health in very different social settings from those found in Western cultures. However, from the rather small number of studies published to date, it seems that these associations are not consistent and straight forward in all settings. Wang (2009) also reports from a study in China that trust at neighbourhood level showed complex relationships with mental health. Harpham et al. (2004) found that in Cali, Columbia, income poverty seemed more strongly associated with distress than reported social capital.

Meanwhile, further research on communities in Britain and North America continues to expose the complex and variable links between social capital and mental health and the different methods used give a diverse set of perspectives on the characteristics of social environments and their variability from place to place, as well as shedding light on the connections to mental health.

Some research presents social environments as structures of social networks which are concentrated in certain areas but do not include everyone living in a place. Fowler and Christakis (2009) use network analysis techniques to explore the idea that emotional states such as 'happiness' can be transferred from one person to another and spread through social networks by a process of 'contagion'. Data from the Framingham study in the USA yielded information on a large number of people who at some point had been included in this longitudinal and multi-generational sample. Their emotional state was measured in the survey, as well as information about the people in their social network. Fowler and Christakis produced computer generated diagrams to visualise the social network of people recorded in the survey (Figure 4.4a). The diagram illustrates the social 'distance' between the members of this network, with those who were closest socially being nearest to each other in the diagram. The analysis also tested the degree of clustering of people who were more or less 'happy' on the survey measure of mental health.

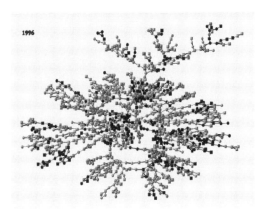

4.4a   Diagram representing a social network

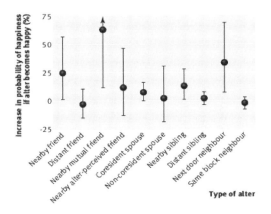

4.4b   Graph showing how happiness for an 'ego' (sample member) related to happiness of social contacts (alter)

Figure 4.4   Social networks in the Framingham survey sample and the relationship with individuals' happiness

Source: Fowler and Christakis (2008), Figures 1 and 4. Reproduced with permission from BMJ Publishing Group Ltd.

Note: Fowler and Christakis explain 'Alter type and happiness in the Framingham social network. Friends, spouses, siblings, and neighbours significantly influence happiness, but only if they live close to ego. Effects estimated with generalised estimating equation logit models of happiness on several different subsamples of network'.

Figure 4.4b reproduces findings by Fowler and Christakis (2009) concerning the association between happiness of each individual sample member ('*ego*') and that of each other person ('*alter*') with whom the 'ego' had a social connection. An *ego* was more likely to be happy if they had a close social connection with a happy *alter*. Also, a person was more likely to be happy if they had friends living in close geographical proximity who were happy. The geographical distribution of happiness was not related to the distribution of people in different socioeconomic groups, so it was argued that happiness was not a function of their social position. This study suggests that the number of individuals with whom one is closely connected (in a social and a geographical sense), and their mental state is significant for one's own level of happiness. The authors suggest that happiness may 'cascade' through social networks within communities and their account emphasises the significance of socio-geographic proximity for this process.

The study is appealing from a geographical perspective, since it provides a visual impression of the complex intermeshing of social networks among a large group of people and the variability of social connection and of mental health that exists within a group of people over time. Also, the finding that geographical proximity is significant suggests that the social environment depends to some degree on the organisation of social groups in physical space. This type of analysis does not, however, tell us much about how or why happiness may 'spread' from one person to another and, strictly speaking, it does not even provide direct evidence that 'happiness' is transmitted from one person to another through social contact. Another possible explanation is that those in closer contact are most likely to be jointly affected by some other process which affects their mental health. This kind of study therefore needs to be complemented by other research that is more clearly focused on causal pathways.

Other authors have also concluded that 'proximal' social connections for individual people may be most important for mental health, rather than more general measures of the social environment prevailing in a wider community or society more generally. Veenstra's (2005) analysis of social capital data for individuals and neighbourhoods in British Columbia, Canada (see above) concluded that most of the variation in depression was among individuals rather than among neighbourhoods. Veenstra expresses some skepticism as to whether 'ecological' differences in social capital at the community level are really important compared with individual variability that is more likely to relate to the immediate social environment of household, family and personal friendship networks. However, there was some evidence that access to public spaces for social and cultural activities was relevant to differences in depression in the sample. Similarly Islam et al. (2006) used information on municipalities in Sweden linked to information on voting participation as a measure of social engagement relating to social capital. They found that quality of life scores were somewhat better in areas with higher levels of voting participation, but the association only explained a small proportion of the variation in quality of life and much larger differences were observed at the level of individuals than among areas, suggesting that quality of life was not strongly associated with this measure of collective social engagement.

Stafford, De Silva, et al. (2008) reported the results of a British study which linked information from a national health survey with the area indicators of social capital derived using the method described above (Stafford et al., 2003).

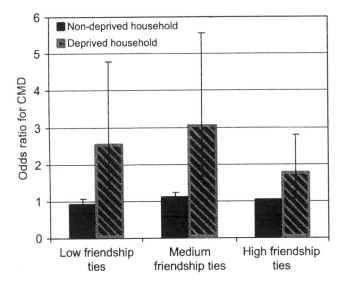

Figure 4.5 **Variation in risk of common mental disorders in relation to individual's household deprivation and the level of friendship ties typical of their residential area**

Note: Odds ratio for common mental disorder by household deprivation and neighbourhood friendship ties. Estimates are adjusted for all other social capital scales, age, sex and social class. Error bars show 95 per cent confidence intervals.

Source: Stafford et al. (2008) Figure 1. Reprinted from *The Lancet* with permission of Elsevier.

This approach is relatively rigorous in that it incorporates detailed information on neighbourhood social capital derived from a survey source that was independent of the information on health, so it avoids the problems of 'same source' bias discussed above. They found no association between area social capital measures and common mental disorders for the health survey respondents as a whole. However, people in lower socio-economic groups living in areas with high levels of friendship ties were less likely to be mentally ill than people who were similarly poor but lived in areas where residents reported weaker friendship ties (see Figure 4.5). The authors therefore suggest that bonding ties between friends may be especially strongly associated with mental health for those living in poverty. This idea is consistent with arguments that social capital may 'buffer' disadvantaged groups from risks to health associated with material poverty.

Some research continues to come from the USA that suggests that the social environment prevailing in a wide area, not only very local conditions, may be important for mental health. For example, Kim and Kawachi (2007) report a study using national survey data from the USA on recent days of poor health due to mental illness. These were linked to state level indicators of social capital. A number of individual variables were controlled for in the analysis which showed that in areas with higher social capital scores people typically reported fewer recent days of mental illness.

Longitudinal research is theoretically more powerful than a cross sectional design if one is trying to demonstrate that the social environment *influences* mental health. For example, Fujiwara and Kawachi (2008) report a study in which mental health at one time point is related to social capital measures from an earlier time point. The measures they used were all derived from the same survey source and related to aspects of cognitive social capital, such as perception of trust in neighbours, and structural aspects such as social contacts and community participation. Perception of trust was associated with lower risk of subsequent mental illness, after controlling for a number of other variables which might also relate to mental health. The strength of this study is the longitudinal perspective on prospective risk, but it also has weaknesses due to the 'same source' bias that might result from measuring mental health and social capital from the same survey, and from a geographical perspective this analysis only measures the nature of relationships in the wider community indirectly, based on individual report. As the authors comment, they were not able to aggregate individual responses within areas to generate ecological variables. This study also investigated the role of previous history of mental disorders as a predictor of mental health at the end of the study period. This accounted for a lot of the prospective risk of mental illness and the authors suggest that social trust may benefit mental health by helping people to *recover* from mental illness rather than by preventing healthy people from becoming ill.

Recently published research focuses on variability within, as well as between communities in the ways that social environments are associated with mental health, showing differences by gender and socio-economic position (e.g. Burke et al. 2009). Research in this vein also focuses on particular groups who have not received as much attention as the general adult population, including children and adolescents. Curtis et al. (2009) reviewed associations between adolescent mental health and the social environment. Drukker et al. (2003) report from the Netherlands that a survey based indicator or informal social control in the neighbourhood was associated with mental health outcomes in children aged 11–12 years, after controlling for their individual attributes. Almedom (2005) comments on this as a relatively rigorous study in the way that it defines the neighbourhood and its social environment. The finding from Maastrict is also supported by Xue et al. (2005) from a study in Chicago USA of children under 12 years. Meltzer et al. (2007) also found in a national study of children aged 11 to 16 in Britain that children's sense of trust in other people in their neighbourhoods showed associations with their psychological wellbeing. Some other examples of the literature focuses particularly on the risks for mental health in areas where social cohesion is relatively weak. This is considered in more detail in Chapter 5.

Some recent studies have emphasised that there may be links between the social environment and 'virtuous' features of the built environment that were considered in Chapter 2. Brown et al. (2009) studied a sample of older people from the Hispanic community in Florida and found that the design of their homes seemed to contribute to their opportunities for social contact and that this in turn was associated with their mental health. Buildings where the frontage was open to the street allowing easier contact with other people passing in the street, which seemed to benefit social interaction for this group. Cohen et al. (2008) showed that proximity to parks in Baltimore, US, was associated with a stronger sense of community efficacy. These findings highlight the connections between the research summarised in this chapter and earlier discussion in Chapter 2 concerning features in the material, physical environment relating to mental health.

Evidence from these statistical studies is complemented by qualitative research which has also revealed associations between mental health, experience of the social environment and processes in social networks. There is a large body of ethnographic research on these questions, much of which considers links to wellbeing broadly defined. For example, Phillipson et al. (2004) edited an interesting collection of studies on social networks and relationships, several of which use intensive and ethnographic methods). However, relatively few explicitly relate their research to mental health outcomes and if they do collect information on health and wellbeing it tends not to be designed to be interpreted in terms of presence or absence of mental illness. This body of work also raises the interesting question of whether we should think of social contact and integration and social support as an aspect of wellbeing and a 'health outcome' in itself, rather than seeking to establish its relationship to the presence or absence of mental illness or distress. This line of argument runs back to discussions referenced above over whether we can view social integration and trust as a 'normative' social good.

Where qualitative research does directly address health and wellbeing agendas, it makes an especially significant contribution by offering insights into the intricate associations between material and social conditions and health and wellbeing. These are not deterministic, but can be quite complex and reciprocal relationships. It also often shows how these relationships play out variably in different settings, which underlines the argument that 'geographies of social relationships' are important to consider in relation to health variation.

For example, the following examples reported by Cattell (2004b: 153) showed informants in her East London study perceived benefits from participation in community based organisations. Involvement in a local residents' association was seen as a factor in the process of regaining a stronger, healthier mental state by one respondent in Cattell's study, who said: "I felt terribly isolated, and had little confidence then…I was on Valium…That part of my health has improved over the last few years, since I've been on my own and involved in the Tenants' Association. I feel in control of my life now, I didn't before." Another resident involved in a local credit union described how this organisation also played a role in providing a 'social hub' where members "…come in for a chat, make friends, become more confident and assertive" and suggested that '…it picks them up a bit". Thus certain spaces in the material neighbourhood environment functioned as sites for collective engagement in activities with both practical and social significance, and helped to 'constitute' and maintain healthy social relations in the community.

Qualitative studies also help us to understand the significance of 'loose' ties among people who share the material space of a local neighbourhood contribute to a sense of belonging in the social space of the local community. Cattell, Dines, et al. (2008, 552–3) discuss respondents' views that social encounters and relationships in their neighbourhood contributed to a positive mood and sense of wellbeing. For example, study participants were quoted as saying:

> When I'm at home I get really stressed with the kids. I'll leave the house and I'm totally stressed but I'll walk round to school, I see a couple of people [on the way], say hello, they smile, and it just all goes. By the time I go back home, I'm a very chilled, different person.

> Because people have such busy lives now…you have to go out of your way to see other people. But if you see people all around your area, you're seeing them day to day, it makes you feel good because you've known them a long time.

Popay and colleagues' mixed method approach to research on health inequality in northern England (see above) generated qualitative as well as statistical information on the role of local social institutions and caring relationships in supporting potentially vulnerable members of the community, as for example in the following quote reported in Popay, Thomas et al. (2003: 67):

> …you're looked after by the neighbours, we have a neighbourhood watch…I've been to meetings in the coordinator's house …we all look after one another …especially the older ones...

Popay, Bennett et al. (2003: 15) also report a similar type of comment from the resident in a socio-economically disadvantaged neighbourhood saying:

> …there's a community spirit…we are all in the home watch...We confide in people and talk to them…It [the sense of community] gives you more confidence, you know. You're not as frightened 'cos you know you can rely on your neighbours…I think trust is one of the big things…If you can trust the people around you then you've got peace of mind…

The significance of these findings about the 'normative', supportive role of community social relationships for people who might be at risk of poor mental and physical heath is also stressed by Day (2008) in her account of older peoples experiences in 'healthy' public spaces in their local neighbourhood. The older respondents she quotes described the opportunities for social interaction as well as the 'emotionally uplifting' effect of green surroundings as attributes of healthy places (which seem to draw on ideas reviewed here in Chapter 2).

These examples all contribute to the impression that the social environment, within the geographical area that one occupies on a daily basis, contributes to one's psychological health and sense of wellbeing. These 'supportive' aspects of the social environment confer resilience to stressors of everyday life. In areas where many people live in relative poverty, or the physical environment is degraded, local residents will nevertheless often emphasise the 'positive' social aspects of their local area, citing what they see as social coherence and strengths of the social relationships they experience with their neighbours and workmates. Wakefield and McMullan (2005) describe how residents of very polluted industrial area of Hamilton, Ontario, Canada tended to describe the value of the strong social ties in their community and to downplay issues relating to the physical environment. Their perceptions contrasted with the views of people from outside the area, who tended to view it as unattractive. Wakefield and McMullan comment that this underlines the varying, socially and geographically contingent ways that therapeutic landscapes are perceived and understood.

The 'ethnic density' effect – social composition as social context

The discussion above has emphasised the potential for 'supportive' neighbourhood social environments to confer psychological resilience in the face of socio-economic disadvantage. These supportive aspects of social environment depend both on the nature of the social setting, and also on the characteristics of the person and the ways that they interact with their environment. This is illustrated, for example, in studies which have investigated the significance of 'social density' or 'congregation' for the health of social groups at risk of disadvantage, discrimination and marginalisation in wider society. Arguments about benefits of 'social density' have been applied extensively, though not exclusively, to studies of the health of minority ethnic groups living in ethnic 'enclaves' in major cities. Ethnic minorities often experience problems of poverty, material deprivation and racism in the host society, and they are frequently concentrated in some of the most deprived and physically degraded parts of cities. However, according to the 'ethnic density' hypothesis, the fact that they are concentrated together may offer some advantages for their health, because geographical proximity helps them to benefit from the support that they gain though bonding ties within their cultural community, conferring advantages of cognitive and structural aspects of social capital. A growing number of studies have investigated this idea empirically by exploring the 'ethnic density' of local populations and the health of individual residents.

Halpern (1993) is often credited for his early commentary on the significance for mental health of minority status in one's community, and arguing that psychological health advantages accrue from living amongst others who belong to the same social group, probably because this facilitates processes in the shared social environment offering both material advantage and social support.

Since Halpern's paper a number of further studies have explored this idea both theoretically and empirically and further reviews have also been published. Wickrama et al. (2005) place the ethnic density hypothesis within a broader argument about 'compound disadvantage' through which individual and neighbourhood risks accumulate for certain ethnic groups and create negative impacts on mental health due to perception of relative disadvantage and status anxiety. They claim that 'race' (i.e. the social significance of race/ethnicity) may moderate the mental health effects of compound disadvantages in three ways. First the *exposure* to cumulative disadvantage may be variable depending on one's race. Second, race may be associated with stronger or weaker levels of *resilience* to community disadvantage. Third, where there is 'incongruent fit' between ethnic identity and the ethnic profile of the local area this may produce added disadvantages for the individual. Ethnic congregation may reduce sense of incongruent fit, and increase resilience through solidarity and social support from others in the group. Pickett and Wilkinson (2008) place particular emphasis on differences in perceived social status of ethnic minorities in communities where they are more or less congregated. It seems debatable whether this status anxiety issue is the sole driver producing health advantages of ethnic density, but it may be consistent with varying experience of racism for people from minority groups. Findings by Karlsen and Nazroo (2000) show that experience of racism was the aspect of minority ethnicity most strongly associated with health outcomes. This observation may help to explain the better health of minorities living in ethnic enclaves where they are less likely to be subject to racist attitudes. It is further supported by findings by Hunt et al. (2007) who report that experiences of racism among Black women in the USA were less common for those living in areas with a higher concentration of other Black residents. Nolh et al. (2007) also found experience of non-overt racism was significant in the development of depression.

Whitley et al. (2006) carried out a qualitative study of what might be considered as the 'inverse' of ethnic density effects in the Gospel Oak area of London, which has relatively low concentrations of ethnic minority groups. Their respondents reported disadvantages including: exclusion from social networks of others in their ethnic group; low levels of provision in their residential area of culturally specific services that they would have liked to use; a perception that the risk of racial abuse was relatively high in their neighbourhood, where they constituted a minority, and a sense of intimidation by the majority population. This study, suggests that the advantages of ethnic density are not limited to questions of relative status, but also relate to other aspects of social and cultural capital including a sense of cultural affinity with others and access to social resources more broadly. Yuan (2008) also found in a study of Black and Hispanic people living in Illinois, USA that social support was reported to be greater for those in areas where their own ethnic group was more concentrated.

Other empirical studies that have explored the question of 'ethnic density in relation to mental health include one by Halpern and Nazroo (2000), based on a national sample of people from minority groups living in Britain, which found negative associations between risk of psychiatric symptoms and ethnic density, although the strength of association was modest, and not consistent across groups (for example, it was reversed for Pakistanis). Morgan and Fearon (2007) review results from the *AESOP* study in the UK and conclude that neighbourhood factors such as ethnic density may be important for psychotic disorders. Veling et al. (2008) reported from a study in the Hague, Netherlands, that while there was some evidence that psychotic disorders were more common among immigrant groups compared with the 'host' population, the differences were only significant when they lived in areas of low concentration of minority groups. Also, Kirkbride et al. (2007) found some evidence in London, UK, that risk of Schizophrenia for individuals from ethnic minorities was lower in areas with relatively large ethnic minority populations. However, the association was only marginally significant in statistical terms.

Some studies do not support the idea of an 'ethnic density effect'. Henderson et al. (2005) tested the association between depression and ethnic concentration in a sample aged 28–40 years from several different locations in the USA. They did not find any association with ethnic concentration (however, they did not test the interaction between ethnicity of the individuals and area ethnic concentration).

While much of the evidence relates to adults, there is also evidence that psychological health advantages for minorities associated with ethnic density may develop in childhood. In the USA, Wickrama et al. (2005) found evidence from a sample of 15,500 young people that adolescent black individuals have better health when located in areas with high concentration of other black people, and this association was significant after controlling for individual and family disadvantage. In subsequent publications (Wickrama et al., 2008) they have emphasised this life course development of social and ethnic difference in mental health which emerges early in life. These findings are echoed in a study from the UK, by Fagg et al. (2006) who linked census data on ethnic composition in small areas to individual data on adolescents in local schools living in a deprived area of East London. Among 'South Asians' (whose families originated in the Indian subcontinent) there was a significant association between the risk of distress reported by individual adolescents and the level of concentration of their own ethnic group in their residential area (after controlling for a number of other individual socio-demographic variables). The diagram in Figure 4.6 shows that the association was 'U' shaped, rather than a constant, linear slope. This suggests that the association between ethnic density and mental health outcomes may be complex and that while moderate levels of concentration are associated with the best levels of psychological health, very high ethnic densities may not show the same link to better health.

Other researchers have reported similar findings of a 'U'-shaped link between ethnic density and risk of mental illness. For example, Neeleman et al. (2001) report this pattern of association between ethnic density and self harm in adult populations in South London, and also found an association between suicide and ethnic concentration (Neeleman and Wesseley, 1999). French (2009) reports a study in Northern Ireland where higher levels of segregation of Protestants and Catholics within some small areas was associated with relatively high levels of consumption of

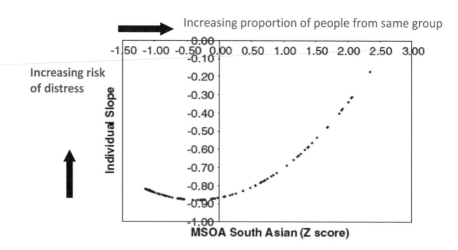

Figure 4.6   Mental health related to 'ethnic density' for East Asian school children in East London

Note: East Asian children have relatively good mental health (less distress measured by the strengths and difficulties questionnaire), especially in areas of moderate 'ethnic density' (after allowing for other individual characteristics.

Source: Fagg et al. (2006), Figure 1. Reprinted from *The Lancet* with permission from Elsevier.

medications and treatment for anxiety and depression by the residents of these areas, after controlling for socio-economic indicators reflecting relative poverty levels.

It therefore seems possible that moderate levels of social congregation and segregation may be helpful for mental health, probably fostering 'bonding ties between people of the same group and strengthening group solidarity. However, extreme levels of ethnic density seem less beneficial. This may be because extreme congregation results in isolation and segregation from the rest of the community which weakens the 'loose', 'bridging' ties with people from other social groups. As noted above the latter are beneficial for mental health because they contribute to social inclusion and access to resources in wider society.

More generally research on the 'ethnic density' effect raises some interesting questions about what we understand by the idea of 'socio-geographical context' introduced in Chapter 1 of this book. In some respects, 'ethnic density' could be considered as a compositional feature of a setting, resulting from the ethnic characteristics of the individual people congregated in an area. On the other hand, 'ethnic density' is considered to act as a 'contextual' attribute of the wider setting because the congregation of individuals from the same group creates a social space with certain attributes that influence the experiences of many people in the area. This example therefore underlines the difficulty of making a clear distinction between 'compositional' and 'contextual' 'effects' on mental health. It draws us towards a perspective that instead interprets 'composition' and 'context' as closely interwoven systems of actors and networks.

The ethnic density effect' arises from the *combined* effect of individual and community ethnicity. Therefore, the literature on 'ethnic density' illustrates why statistical studies aiming to distinguish statistically *independent* contextual and compositional effects on health often give rather inconclusive results, and may be at odds with the perceptions of the importance of the collective social environment, reported in qualitative studies. In statistical research the *interactions* between individual and area conditions, as they relate to mental health, may therefore be particularly illuminating and deserve more attention.

Conclusion

This chapter has reviewed parts of the research on the associations between mental health and social relationships. Here we have focused especially on social relationships at the level of communities rather than those operating within families. There is considerable theoretical support, and some, more tentative empirical evidence for the argument that strong, cohesive social relationships that help to integrate individuals into their local community and offer access to resources for health may have direct or indirect associations with mental health of individuals.

It has also been noted that from a methodological perspective it can be quite challenging to generate information that allows rigorous assessment of the significance of the neighbourhood social environment for mental health. Nevertheless, considerable progress has been made in the last decade to develop more sophisticated strategies, and these are beginning to provide new insights. It remains the case, however, that some of the most convincing evidence comes from qualitative studies in particular neighbourhoods, showing that local residents perceive aspects of social capital and social cohesion to be beneficial for their state of mind. Both quantitative and qualitative studies reported here suggest quite complex associations may be operating which vary from one place to another and also among individuals living in the same place. One's experience of the local social environment depends on shared experiences and social interactions built up over time. It also seems likely that there are reciprocal links between one's mental health and one's social relationships, which have mutual effects on each other.

So far the discussion has mainly focused on aspects of social environments that are often considered to be benign for mental health. However, we have also noted that the same processes can contribute to inequalities of health within and between different local communities. In Chapter 5 this discussion is extended to consider aspects of the social environment that are detrimental to public mental health.

It is interesting from a geographical perspective that spatial proximity is important for the operation of social ties that help to create a 'healthy' social environment. This does not mean however that one's social world is circumscribed by contacts in one's immediate neighbourhood, and consideration should also be given to wider 'virtual' and 'imagined' communities that do not necessarily involve spatial proximity and physical contact among their members. For the arguments in this book it is also important to underline that the social environment interacts with the material and symbolic environment in quite complex ways. Thus it is apparent that there are connections between the ideas in this chapter and those in Chapters 2 and 6 of this book concerning the material and the symbolic landscapes associated with mental health.

Learning objectives

Material in this chapter aims to help students to:

- understand theories of the nature of 'social capital' and why and how it is associated with mental health;
- understand how social environment can be assessed at the community level, using statistical and qualitative approaches;
- understand why a distinction between social 'composition' and social 'context' is difficult to make and the importance of being able to see how individual social attributes interact with community characteristics in ways that are related to mental health.

Introductory reading

Well articulated conceptual frameworks linking social environments to psychological health are offered by:

Berkman, L.F., T. Glass, et al. (2000). From social integration to health: Durkheim in the new millennium. *Social Science & Medicine*, 51(6): 843–57.

Cattell, V. (2001). Poor people, poor places, and poor health: the mediating role of social networks and social capital. *Social Science & Medicine*, 52(10): 1501–16.

Recent reviews with interesting discussion of the different dimensions of social capital as are provided by:

Fulkerson, G.M. and G.H. Thompson (2008). The Evolution of a Contested Concept: A Meta-Analysis of Social Capital Definitions and Trends (1988–2006). *Sociological Inquiry*, 78(4): 536–57.

De Silva, M.J., K. McKenzie, et al. (2005). Social capital and mental illness: a systematic review. *Journal of Epidemiology and Community Health*, 59(8): 619–27.

Research on the 'group density effect' has recently been reviewed by:

Pickett, K.E. and R.G. Wilkinson (2008). People like us: ethnic group density effects on health. *Ethnicity & Health*, 13(4): 321–34.

Chapter 5
Anomie, Status Anxiety and Fear: Socio-spatial Relations and Mental Ill-health

Summary

This chapter extends the discussion of the social environment in Chapter 4, by reviewing research on aspects of the social environment associated more particularly with risks of mental illness. Taking a geographical perspective, theories and measures of social fragmentation and relative poverty are considered in terms of their association with aspects of mental health. Social indicators for geographical areas are shown to relate to variation in indicators of severe mental illness, such as suicide and schizophrenia, as well as measures of less severe distress. This chapter considers the connections between degradation of the material environment and social incivility and stress, sometimes referred to as the 'broken window phenomenon'. It explores arguments relating to the psychosocial significance of relative poverty and inequality at the community level. Attention is also given in this chapter to issues of reverse causation, and selective migration which complicate the interpretation of the links between the social environment and mental health.

Introduction: The *anomie* hypothesis: social fragmentation and psychological distress

In this chapter the focus is on aspects of the social environment which may be damaging to one's psychological state and present risks for mental illness. The discussion focuses especially on ideas relating to the 'anomie' hypothesis (drawing from work originating with Durkheim, concerned with the impact of modern living on our sense of integration in wider society), and ideas about relative poverty and 'misery' which explore the relationships between socio-economic position in society and psychological health. As in Chapter 4, the discussion here is especially focused on social relations at the 'ecological' level of communities, and whether these interact with individual social attributes, so the emphasis is on the processes in collective social space that may relate to mental illness.

As discussed in the previous chapter, there is a well established discourse in the social and medical sciences concerning the importance of social relations for mental health. Several commentators have argued that trends in mental health have been historically and geographically associated with changes in society and

Box 5.1   Durkheim's theory of *anomie*

Durkheim first published his theories about anomie and suicide in 1897. He argued that a 'healthy' society should play a 'moderating role', placing 'just limits' on expectations through processes of 'moral education'. He was seeking to understand why, although people in French society were on average wealthier and lived more comfortable lives than ever before, suicide was apparently increasingly common. He developed a theory that the sudden changes in wealth seen during the industrial revolution disturbed socially determined aspirations and social relationships, so that modern societies were no longer able to provide this 'moral education' on the basis of traditional values. Instead there was a growing tendency to promote purely economic goals and ever-increasing expectations and desires for material wealth. (A tendency now encapsulated in popular representations of the 'consumer society'.) As a result, people became anxious about attaining increasing material improvements to their lifestyle, always seeming beyond their reach, and any set-back in economic advancement would be very distressing. For example, Durkheim described modern societies as follows:

> ....nations are declared to have the single chief purpose of achieving industrial prosperity...industry, instead of being still regarded as a means to an end...has become the supreme end of individuals and societies alike...Thereupon the appetites thus excited have become freed of any limiting authority...From top to bottom of the [social] ladder, greed is aroused without knowing where to find ultimate foothold...A thirst arises for novelties, unfamiliar pleasures, nameless sensations, all of which loose their savour once known..Henceforth, one has no strength to endure the least reverse... (Durkheim [1993 edition], pp. 255–6.

Furthermore, Durkheim considered that the division of labour in industrial societies was detrimental for wellbeing because it broke up established patterns of social relationships, producing social and geographical divisions between labourers and their employers, and separating workers from their family for long periods of the day, which he saw as disruptive to family relationships. These processes, according to Durkheim, were responsible for a sense of 'anomie', or loss of a sense of one's value and 'place' in society. Without the reassurance of a sense of worthwhile life goals, except unattainable economic ones, their sense of social 'worth' was being undermined, and, lacking support from kinship and other non-economic social ties, they were more likely to succumb to despair. According to Durkheim, 'Anomic suicide' was increasingly the outcome and he concludes that: "Anomie, therefore, is a regular and specific factor in suicide in our modern societies..." (Durkheim [1993 edition] p. 258.

especially with urbanisation. It is suggested that social ties and social relations in modern society have weakened, eroding the beneficial effects of social capital that were discussed in Chapter 4, and with detrimental implications for mental health. One very influential exponent of this view was Emile Durkheim (1858–1917) a pioneer of French sociology who set out his ideas in various publications including: *Suicide: A Study in Sociology* and *The Division of Labour in Society.* Durkheim noted that suicide rates in countries like France were increasing due to what he viewed as 'pathological' aspects of modern society that went hand-in-hand with the increasing economic and material wealth and the growing complexity of work and everyday life following the industrial revolution. He argued that humans experience potentially insatiable desires, over and above the basic necessities of life, and that the increased availability of wealth and consumer goods was having a destabilising effect on citizens of countries like France in the late 19th century. Because of these changes, they lost their sense of what were reasonable aspirations and experienced a loosening of non-economic social ties, such as extended family bonds that helped them feel secure in belonging to a social group. Durkheim hypothesised a link between this sense of alienation and risk of suicide (see Box 5.1) .

The early 20th century arguments put forward by Durkheim seemed consistent with another widely cited historic study of severe mental illness, published by Faris and Dunham (1939), which compared crude rates of schizophrenia cases per 100,000 population in different areas of the city of Chicago 1922–1931 and noted a concentration of people with schizophrenia in the inner city 'slum' area. (For a detailed discussion see Jones and Moon (1987: 198–201). This study also made explicit connections between socio-economic conditions associated with the modernisation of society, stressful living conditions associated with urban industrial growth, and rapid social change, including high rates of residential mobility experienced in cities such as Chicago in the late 19th and early 20th Centuries.

The conditions of anomie and urban stress that Durkheim, and Faris and Dunham described, are still of concern in contemporary society and they are the subject of ongoing debate and research. The close ties between family members and neighbours may be weakened by the increasing trend towards smaller households, especially people living alone. Increasing geographical mobility, at least for some social groups, results in greater physical separation between people and more pronounced 'churning' of local populations as people move in and out of neighbourhoods at a growing rate. Research continues to explain the processes of 'anomie' and associated conditions of 'social fragmentation' as they relate to poor mental health, and especially to elucidate the causal pathways that may explain the relationships (e.g. Berkman et al. 2000). Early historical accounts produced largely theoretical conceptual models to account for the relatively simple patterns they were describing (such as trends in suicide, or geographical patterns of disease). Contemporary research develops these theories but also aims to test them empirically. Social fragmentation in communities may be associated with: lack of strong and reassuring positive role models and social norms of behaviour (sometimes termed 'collective efficacy'), causing uncertainty and undermining healthy lifestyles; lack of 'solidarity' with others in one's own group, which reduces available emotional support and practical help; increasing levels of crime and social disorder cause fear and insecurity. In short, it is argued (e.g. Putnam, 2000) that these processes are eroding the levels of social capital in communities, which, as shown in Chapter 4 are important for psychological wellbeing.

Further processes generating variation in mental health at the population level, could be summarised broadly under the 'misery' hypothesis, which incorporates a range of theoretical and empirical evidence exploring how economic deprivation and material poverty are associated with poor psychological health. From a theoretical perspective, there seem to be a number of possible reasons why material poverty may damage mental health or exacerbate existing mental illness or distress. Some of these were reviewed in Chapter 3 where is was noted how individual experience of poverty may have a material impact on mental health; poor living conditions such as inadequate, polluted residential environments are depressing, and represent the antithesis of the 'therapeutic' settings that were reviewed in Chapter 2, and they may also have physiological impacts on brain function. Also, material poverty may operate at the level of whole communities, since in areas with high levels of material poverty, communities may lack facilities and other material resources that help to protect mental health. Furthermore, signs of physical deterioration and damage to built structures may engender unease, or even fear and can contribute to the stigma of neighbourhoods with a poor reputation. Kelling and Coles (1996) drew attention to what is often referred to as the 'broken windows' scenario, where vandalism to property, such as broken windows, that is not held in check and repaired quickly, creates an impression that social incivility and disorder are out of control and that the area is 'uncared for', leading to a further escalation of disorder and generating a vicious circle of spiralling degeneration. Sampson et al. (2002) have reviewed evidence that suggests associations between physical degeneration of areas and crime and social disorder, though they make the point that it is difficult to establish the causal pathways that produce these associations and it is not necessarily the case that prevalence of physical degeneration in an area directly provokes greater levels of social disorder, as predicted by the 'broken windows' hypothesis. It is hypothesised that the combination of social disorder and physical degeneration of neighbourhoods has damaging implications for local residents' sense of security and self esteem and for the public image of the area viewed from outside, leading to stigmatisation of the area and its population. Material poverty and disintegrating built infrastructure in an area may also be *indirectly* associated with mental health, because it is often occurs in places where living conditions are blighted by other stressors, such as unsatisfactory working conditions and unemployment which may be associated with higher levels of depression and suicidal tendencies.

We can see a recurring theme in all of these arguments concerning the significance for mental health of the connections between deterioration of the physical and of the social fabric of society. The *social significance* of material poverty may be at least as important psychologically as the physical reality of poor environments. With respect to both mental and physical health inequalities, this argument is also put forward, for example, by Smyth (2008). Theories dating from those of Durkheim to contemporary authors such as Wilkinson (1996) therefore suggest that perceptions of *relative poverty* may be especially important for psychological health. Wilkinson argues that in high income countries today, severe material poverty, though not eradicated, impacts smaller proportions of the population now than in the past, and the percentage in poverty is much less than in low income countries. Although material poverty is still significant for many people, Wilkinson argues that the psychological effects of relative deprivation are as important for health as the material impacts of poverty. 'Psychosocial factors' have therefore become very important for health. According to this perspective, the level of economic inequality in a country is at least as significant as the absolute average wealth of the population, partly because it influences the *perceived* risk of experiencing poverty for all members of society and accentuates the relative disadvantage of those who are less fortunate, compared with those who are better off. Also, income inequality is thought to be damaging for everyone in society, not only for those who are poor. Wilkinson (1996: 215) states:

> In terms of the quality of life, which is ultimately a matter of people's subjective sense of well-being, the psychosocial processes round inequality, social cohesion and its effect on health, are over-whelmingly important...not only from the point of view of those low down the social scale...but also because the deterioration of public life, the loss of a sense of community, and particularly the increase in crime and violence [which Wilkinson postulates are linked to inequality] are fundamentally important to the quality of life for everyone.

Wilkinson's ideas are of particular interest to health geographers because he seems to suggest that inequality is a 'contextual' factor which influences all the individuals in a particular social setting, and is not only a matter of individual social position. He is also arguing that greater social inequality tends to undermine social capital of the type that was seen in the Chapter 4 of this book to be potentially significant for mental health.

In a related argument, De Botton (2004: 3–4) describes the psychological effects of relative poverty in terms of 'status anxiety', which he defines as:

> A worry, so pernicious as to be capable of ruining extended stretches of our lives, that we are in danger of failing to conform to the ideas of success laid down by our society and that we may as a result be stripped of dignity and respect; a worry that we are currently occupying too modest a rung or are about to fall to a lower one.

One may therefore suffer distress and 'humiliation' especially if one's standard of living is felt to be much worse by comparison with other people in the society, social group or local community that one aspires to. This anxiety, according to De Botton (2004: 4) is provoked by factors including "...recession, redundancy, promotions, retirement...newspaper profiles of the prominent and the greater success of friends." A geographical extension of this argument would suggest that, people assess their own position in relation to their socially constructed vision of the norms and expectations typical in a particular social context, and are influenced by the processes in the socio-economic environment such as economic and social trends (e.g. Smith, 1977, 1994). Furthermore, a stigmatised view held by society of certain places with a poor reputation is likely to extend to the individual people living in an area, in ways which are damaging for their self esteem and psychological wellbeing.

Figure 5.1 summarises the broad patterns of associations that one might expect to find, based on these ideas, with worse health for individuals associated with more deprived social and material environments. The overlap of material and social conditions is intended to indicate the interconnections between social and material aspects of deprivation which are elaborated below. In the following discussion, we explore some of the recent empirical evidence relating to these hypothesised relationships, focusing especially on research concerning the significance for mental health of associations between 'poor' people and 'poor' places.

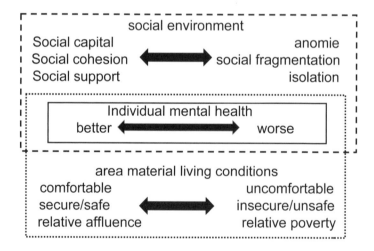

Figure 5.1   Variation in social environment and material living conditions in relation to population mental health

Box 5.2   Designing 'ecological' research on mental health and socio-economic conditions

In order to discover whether local socio-economic conditions in small areas are associated with mental health of the population, the basic considerations for a rigorous research include the following:

- use valid and comparable measures of mental illness in the population standardised as appropriate for relevant demographic factors;
- use theoretically justified area measures of social conditions which can be compared across areas (e.g. statistically standardised area deprivation scores);
- use geographical boundary definitions which correspond well to local differences in socio-economic conditions;
- consider the time data were collected; in general, socio-economic data should be for the same period as the mental health data, however, if, for example, one is exploring the idea that socio-economic conditions *cause* mental illness it may be more appropriate to consider information on social conditions collected *before* the mental health data;
- any interpretation of ecological relationships should consider the 'ecological fallacy';
- interpretations should also consider the possibility of selective migration effects, especially in cross sectional studies.

More complex issues which are also important for rigorous research include these:

- consider the impacts of small counts which may affect data on mental health and social conditions in small area populations – if necessary use Bayesian methods to 'smooth' the data (see Box 5.3);
- consider whether small areas with specific characteristics are clustered together – if so this may indicate spatial autocorrelation which needs to be controlled for, or perhaps a multi-level model would be preferable to test for ecological patterns at different geographical scales;
- if appropriate, use data on individuals within areas to design 'multi-level' analyses which can test the more complex associations between individual's mental health, their socio-economic characteristics and wider environmental factors.

Geographical evidence for the relationships between mental illness, 'anomie' and misery

Since Durkheim and Faris and Dunham (1939) published their findings, an extensive literature on the relationships between mental health outcomes and socio-economic conditions has developed. This long-running discussion concerning geographical associations between socio-economic conditions and mental health has continued to draw on socio-geographical evidence of 'ecological' associations between population mental health and socio-economic conditions in geographical areas. Over time, approaches to this type of ecological study have become more sophisticated and Box 5.2 summarises some of the considerations now thought to be important for rigorous design of ecological research (see also Curtis and Cummins, 2007).

A first step in this type of analysis is to measure how mental health varies across areas. As discussed in Chapter 1, a number of different methods are possible, for example, using data from community surveys on psychological morbidity and distress, routinely recorded data on psychiatric service use and data from deaths registers on mortality due psychological illness and distress, notably suicide. For example, extending the debate about suicide provoked by Durkheim, a number of studies have examined variation in suicide rates at the small area level and investigated their association with socio-economic variation. These include analyses by Saunderson and Langford (1996) of the geographical distribution of age and sex standardised mortality ratios due to suicide for districts of England and Wales, over the period 1989–1992. This was repeated at a finer scale by Middleton et al. (2008) using data for electoral wards for the period 1988 to 1994. Both these studies used Bayesian smoothing methods to produce a more informative map of the underlying pattern of suicide rates (See Box 5.3). These analyses showed that suicide levels are comparatively high in inner city areas, but also in more sparsely populated rural areas. The concentration of high rates in inner cities seems broadly consistent with Durkheim's arguments about modernity and urban-industrial lifestyles as a contributor to malaise. However, rural areas, might be considered likely to be typified by more traditional, close knit communities, so the relatively high rates found there might seem to put Durkheim's interpretation into question. It is therefore important to explore the social and economic factors associated with these geographical patterns.

Research aiming to explain these geographical variations in suicide rates has also used ecological analyses that test for associations between suicide rates and social and economic indicators chosen to reflect conditions of anomie and relative deprivation. Many empirical studies have examined these relationships and Rehkopf and Buka (2006) summarised findings from 86 papers, on 221 analyses from around the world. Their review showed that findings from these different studies varied: in 55 per cent of studies there was *no association* between poverty and suicide. Of those studies which did show a significant relationship, the majority (70 per cent) suggest a positive association such that where poverty is worse, suicide rates are higher. In contrast, 30 per cent of the significant associations

**Box 5.3   Using Bayesian smoothing methods to adjust indicators of
             rates of psychiatric disorders in small geographical areas**

Suicide is, fortunately, a relatively rare event, so at the small area level the numbers of cases of suicide over a period of a year or so may be few or non-existent. This makes it difficult to produce informative maps of suicide over a limited period without adjusting the data to 'smooth' the effects of random variability in small counts and to estimate underlying longer term rates for areas where no deaths have occurred. (Similar problems apply to other, relatively rare, severe psychiatric disorders.) Bayesian 'smoothing' techniques 'borrow strength' from information on the prevailing pattern over other areas included in the dataset in order to estimate the likely underlying pattern of suicides. In areas where the actual number of 'observed' suicides is very small or zero, these estimates based on the larger dataset carry more weight. For areas where suicides are more numerous the actual, observed number of suicides will be more influential in determination of the 'smoothed' rate of suicide. These smoothing techniques need to be applied with caution since they depend on the 'prior' assumptions built into the calculations, and as with all statistical estimates there is a margin of possible error, which can also be calculated. Also, as Middleton et al. (2008) point out, the effect of 'smoothing towards the mean' for data on very small areas can be to 'loose' the impression of geographical variations which may be theoretically, if not statistically important. Saunderson and Langford (1996) and Middleton et al. (2008) used this Bayesian smoothing technique to calculate rates of suicide in England and Wales. See Figure 5.2 illustrating the effect of smoothing on the distribution of district Standardised Mortality Ratios calculated by Saunderson and Langford (1996). Middleton and colleagues illustrate maps produced with and without smoothing, to demonstrate the effect of the procedure. These maps show that rates of suicide are relatively high in rural areas with very small populations, and small absolute numbers of deaths, as well as in some inner city areas, where the population in each small area is larger.

were in the reverse direction, that is, suicide was higher in areas with better socio-economic conditions. The pattern of the findings varied according to the *scale* of the areas used in the study: the finding of a positive association between suicide rates and poverty was more common in studies of small areas, as opposed to large regions or international differences between countries. Positive associations were also more likely where the measures of socio-economic conditions related to poverty measures like unemployment or other aspects of material deprivation (fewer studies showed this association when using measures of average income

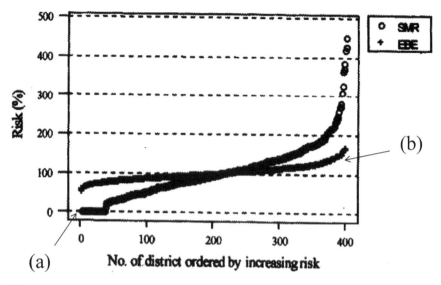

Figure 5.2   **Graph showing the effect of Bayesian smoothing on age standardised suicide ratios**

Note: Plot line (a) shows the distribution of the unadjusted SMR for suicide among males aged 15–24 in local Government districts in England and Wales. It represents without adjustment areas with zero values where no suicides occurred during the study period and extremely high rates in some areas. Plot line (b) shows the 'smoothed' estimated mortality ratios after applying Bayesian methods. There are no zero values and extremely high rates are estimated as closer to the average.

Source: Saunderson and Langford (1996). Reprinted from *The Lancet* with permission of Elsevier.

that were not translated into indicators expressing standard of living). Their review therefore seems to suggest that deprivation and poverty (rather than the level of income level as such) may be most significant for mental health, and seems consistent with arguments that relative poverty and material deprivation, rather than absolute income levels, are more relevant for mental health.

As suggested in Box 5.2 basic methodological considerations in ecological research include the measures of mental health and social and economic factors that are employed and various studies have used different strategies for defining socio-economic conditions. The area indicators should be selected to reflect available theoretical and empirical information concerning the aspects of disadvantage that are likely to be associated with mental health, including both material poverty and social fragmentation. Small area indicators of material deprivation and socio-economic disadvantage have often been compiled using population census data. For example, for areas in England and Wales, the 'Townsend Score' of material deprivation (Townsend et al., 1988) has been widely used in research on health inequalities, and is useful because it clearly focuses on material, as opposed to social aspects of disadvantage. Different dimensions of disadvantage have often been shown in research to be independently associated with mental distress and psychiatric morbidity, and they often have different geographical distributions. For example, Congdon (1996, 1997) and Whitley et al. (1999) used the Townsend score as well as a separate measure of 'social fragmentation' to assess the relationship with local rates of suicide in Britain. They showed that prediction of suicide rates could be improved by using both measures (see Box 5.4). This seems to suggest that social relationships as well as material poverty may combine to create environments associated with poor mental health as anticipated in Figure 5.1.

In the USA attention seems to have focused more on neighbourhood factors linked to social capital (see Chapter 4) and social disorder (e.g. the 'broken windows' debate considered below). However, the potential of area data to clarify the role of neighbourhood socio-economic deprivation in relation to health outcomes such as mental illness has also been considered (e.g. Eibner and Sturm, 2004) and some authors (e.g. Subramanian Chen, Rehkopf, et al., 2005; Subramanian, Jones and Kaddour, 2009a, 2009b) now argue that it is essential to consider area poverty measures in order to fully understand the inequalities in general health experienced by social groups defined by race and ethnicity.

The disadvantage of census based measures of socio-economic conditions is that they can only be updated when a new population census is carried out. In England, the Index of Multiple Deprivation (IMD) developed by Noble et al. (2004) and disseminated by the government for use in research and planning, incorporates routinely available data which can be updated more frequently. The IMD includes information on a range of aspects of deprivation, with subscales relating to: housing, education, employment, income, general health and disability, access to services, social environment. Studies in England have used area IMD measures to analyse their association with mental illness. For example, Rezaeian, Dunn, and St Leger (2005), have demonstrated that this measure shows an ecological association with the risk of suicide, especially for younger and middle aged men in England. This indicator has also been used in other studies discussed later.

One of the limitations of the type of ecological analysis of small area data shown in Box 5.4 is referred to as the 'ecological fallacy'. The associations applicable to populations grouped in areas do not necessarily apply at the level of individuals. (For example, if *populations* in poor areas tend to have worse health than in wealthier areas, this does not prove that it is the poor *individuals* within these areas who have worse health than wealthier people). Partly to address this issue, ecological data for areas may be combined with information for individuals from surveys or other sources, to investigate whether individual risks for mental illness relate to the type of setting where the person lives, as well as the individual's own characteristics. This approach sometimes uses multi-level models to test associations for individuals clustered within areas. (For reviews of recent applications of these models in health research more

Box 5.4   Ecological associations between suicide, material deprivation and social fragmentation in Britain

Examples of research relating material/economic deprivation and social fragmentation to mental health outcomes include two studies: one on suicide rates in London by Congdon (1996, 1997) and the other, a national study for 633 parliamentary constituencies in Britain by Whitley et al. (1999).

Both studies used Townsend's material deprivation score for the geographical units examined. Townsend et al. (1988) proposed combining the following census variables to calculate a composite score of material deprivation for small areas. These indicators might denote factors which have some direct relevance for health, but they were really intended as proxy measures of more complex ideas about material poverty (shown in parentheses in the following list):

% of working age population unemployed
(precarity in labour market);

% of population in housing which is not owner occupied
('housing wealth'/long term income);

% households in overcrowded housing
(poor living conditions);

% of households lacking a car
(disposable income/ownership of consumer durables important for lifestyle).

Congdon (1996) used Townsend's measure and he also compiled a separate measure based on census data for small areas to create a 'social fragmentation' score for small areas. This combined information on the following list of simple variables, that Congdon argued would be associated with low social cohesion, lack of traditional structures supportive of social capital and greater probability of social disorganisation in the local community as a whole (shown in parentheses).

% living alone
(lack of family social support within the home and general social isolation);

% adults not living in married couples
(non-traditional social relationships);

% of population who moved home in the year prior to the census)
(instability in the composition of the area and 'churning' in the
population);

% in rented accommodation (those living in rented accommodation may
have less long term 'commitment' to the community than those who have
invested in owner occupied housing).

Congdon (1996, 1997) analysed suicide in boroughs and in electoral wards
in London for time periods around 1990. He used conventional Poisson
regression models as well as Bayesian modelling to test how suicide rates
related to these measures of deprivation and social fragmentation. Male
suicide rate was predicted by material deprivation but including information
on social fragmentation did not improve the predictive power of the model.
The social fragmentation score was more strongly associated with female
suicide rates than the deprivation score. Thus social factors may relate to
risk of suicide in different ways for men and women. Congdon (1997) also
included in his models information about how small areas with similar rates
were clustered in space and showed that there may have been other factors
operating across groups of areas which partly accounted for the pattern
of suicide. The associations between socio-economic conditions and ward
level suicide rates varied across different parts of London, suggesting that
the significance for suicide of social conditions is variable, depending on
the local context.

Whitley et al. (1999) carried out research on variation in suicide in 633
parliamentary constituencies in Britain. They used population weighted
least squares regression techniques to analyse the association of suicide
rates during the 1980s with Townsend's deprivation score and with
Congdon's social fragmentation score. In this national study, they found
that the deprivation and social fragmentation scores both had independent
associations with suicide, especially for younger adults under 65 years of
age. Allowing for this, they also found that higher suicide rates occurred in
areas where social fragmentation had increased over the 1980s. This paper
therefore supports Congdon's finding that social fragmentation as well as
material deprivation may be important for deprivation.

generally see: Pickett and Pearl (2001) and Riva et al. (2007)). Several of these studies also use area measures of both material and social deprivation, similar to the indices developed by Townsend and Congdon, and they include research from countries outside the UK.

For example, in Canada, small area indicators of social and material deprivation were developed by Pampalon et al. (2004), which are very similar to the British Townsend and Congdon indices. Their *material* deprivation indicator comprises information on the population aged over 15 years: proportions without a high school certificate or diploma; proportions unemployed and average income. It is meant to reflect average levels of financial and economic poverty in the local population. Their *social* deprivation measure comprises data on the proportion of the population over 15 years old who live alone and the proportion separated, divorced or widowed, as well as the proportion of households that are single parent families. Curtis, Setia and Quesnel-Vallée (2009) combined these indicators with individual data from the *Canadian National Population Health Survey*. Data on individual characteristics recorded in the survey were linked with information on the category of social and of material deprivation in the area where the person was living. Although people living in the most socially deprived areas had significantly worse health than those in the least deprived areas, much of the association between area social deprivation and individual distress seemed to be explained by individual variables denoting socio-economic and demographic attributes of the person (Curtis, Setia and Quesnel-Vallée, 2009).

A second example of a census based measure of area conditions, from the USA, is reported by Silver et al. (2002) who studied small area variations using an indicator of social disadvantage that comprised: proportions of small area populations receiving public assistance payments, proportions of households that were husband and wife families, proportions of female headed households with children, persons below the poverty line, adult unemployment rate, adults in executive or managerial jobs, families with high annual incomes. They also used a mobility index comprising relative numbers of people moving in the 5 years prior to the census and the proportions in rented housing. They combined these indicators with survey data for populations in five major US cities and found that, controlling for individual risk factors for mental illness, people living in more 'disadvantaged' areas had a higher risk of major depression and that high levels of neighbourhood residential mobility were associated with risks of major depression and of schizophrenia. This finding therefore supports the argument that the wider social context of neighbourhood social disadvantage relates to risk of depression as well as individual characteristics.

Another strategy employed in a number of studies is to investigate the importance of area poverty and social fragmentation using survey data from a geographically clustered sample of individuals giving information on mental health and also on their views of material deprivation or social fragmentation in their area. Individual views of area conditions might be influenced by their mental health (e.g. see Fagg, Curtis, Clark et al., 2008, discussed below), so in more rigorous studies the measure of area conditions is produced by aggregating responses for all the people surveyed in each area (see Chapter 4). In some cases, too, independent observation by trained observers is used to assess visible material deprivation in areas. For example, Fone et al. (2007) report a study using data on over 10,000 respondents in Wales, which aggregated survey responses by residential area and used factor analysis to generate a social cohesion subscale for areas. People living in areas with the lowest social cohesion score were found to have significantly greater average scores of mental distress. A measure of area income deprivation for each respondent's area of residence was also found to be positively associated with mental distress, after controlling for individual variables.

These examples show that the association of area deprivation with mental health of individuals can be difficult to distinguish from the hardships they face at the individual and family level, and that findings on this question vary between studies. Mair, Diez Roux and Galea (2008) reviewed 25 papers which considered the associations between neighbourhood socio-economic conditions and depression, using research designs which involved information on areas and on individual mental health and controlled for individual risk factors when testing area variables. They found that only 13 studies reported that, after controlling for individual socio-economic variables, there was an association with area differences. The lack of consistency in research findings on this question is probably not surprising in view of the variability in measures used, though it is also possible that community factors have stronger impacts on mental health in some places than in others. This might occur, for example, if the socio-economic inequalities between neighbourhoods are more pronounced in some regions than in others.

Some of the studies in this field are of particular interest because they use a longitudinal design which is theoretically more powerful than a cross sectional approach. For example, an American study by Yen and Kaplan (1999) used information from the Alameda County study to assess risk of depressive symptoms. They showed that people in poor areas were at higher risk of developing depressive symptoms, but that these small areas differences in risk seemed to be explained by their individual characteristics. Similarly, Weich, Twigg, Lewis et al. (2006) combined longitudinal information on individuals' mental health and socio-economic characteristics from the *British Household Panel Survey* with a census-based socioeconomic indicator for electoral wards proposed by Carstairs (1990), similar to the Townsend index described above. They used multi-level modelling to test how risk of developing common mental disorders related to individual, household and area conditions. People living in more deprived areas were more likely to develop common mental disorders, but this greater risk seemed to be mainly associated with their individual or household characteristics, rather than with conditions in their neighbourhood. Curtis, Setia and Quesnel-Vallée (2009) in their study of a national longitudinal survey sample in Canada, also found that social or material deprivation in place of residence at one time point did not predict subsequent variations in development of psychological distress at a later date when individual predictors were included in the analysis.

While it is not possible to draw general conclusions from these studies alone, they do underline the importance of considering prospective risk of development mental illness, rather than using a cross sectional design in order to address possible effects such as selective migration or pre-existing mental illness, discussed below. Also, the three studies cited here were well designed and relatively rigorous because they controlled for individual attributes. The authors seem to agree that differences in individual characteristics are more powerful predictors of mental health than area conditions, although there may be complex interactions between individual and area attributes which their studies did not identify. These studies do, however, all show that socio-economic disadvantage predicts subsequent emergence of mental health problems later in the life course, which is consistent with the idea that 'misery' might directly or indirectly 'cause' psychological morbidity. This raises the question of how early in life personal socio-economic differences in mental illness may arise and whether area deprivation may play a role in mental health inequalities among young people as well as older populations.

The example of adolescent mental health and area disadvantage

Many of the published research studies on area deprivation and mental health relate to the general adult population. However, mental health problems can develop relatively early in the life cycle compared with many physical illnesses, and there is concern in some countries that levels of psychiatric disorder may be increasing among young people (e.g. Collishaw et al., 2004). It is therefore interesting to consider some illustrations of research on area disadvantage which is focused on younger populations.

A study of over 5,000 young people aged 16–24 in England, by Fagg, Curtis, Stansfeld et al. (2008) linked national survey data on individuals to information about the levels of social fragmentation and material deprivation in their area of residence. They showed that young people living in areas with higher scores on Congdon's social fragmentation index were more likely to report distress (see Figure 5.3), and the association was significant after controlling for a range of individual social variables. This study also investigated whether the IMD measure of material deprivation would predict mental distress in this sample, but this area measure was not significantly predictive of variation in mental health when measures of individual socio-economic position. were included in the statistical model. This finding concerning material deprivation seems to be corroborated by Ford et al. (2004) using national data for a younger English sample from the *British Child and Adolescent Mental Health Survey* and an area deprivation indicator based on Carstairs' (1990) deprivation index. Although children living in more deprived areas, ranked by the 'Carstairs index' did have higher prevalence of several types of psychiatric disorder, the differences were almost entirely accounted for by their individual characteristics and there were no 'independent' associations with area conditions. Taken together these two studies suggest that

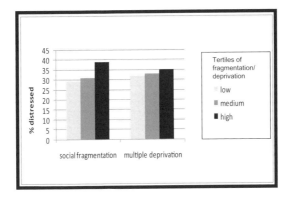

Figure 5.3   Young people in England: trend in per cent 'distressed' by social conditions in place of residence

Source: Fagg, Curtis, Stansfeld et al. (2008). Reproduced with permission.

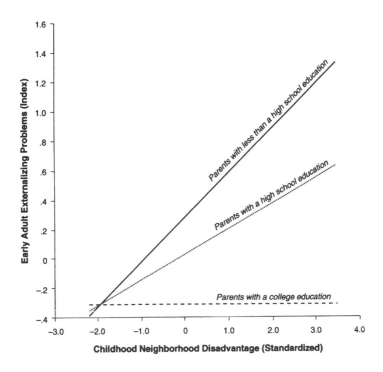

Figure 5.4   Relationships between externalising psychological problems in early adulthood, neighbourhood deprivation earlier in life and parental education

Source Wheaton and Clarke, 2003 p. 696. Reprinted with permission of author and publisher.

for young adults in England social cohesion/fragmentation in the community may be important for mental health, but material deprivation at area level is not, when individual poverty is taken into account.

The findings from these two studies of young people from England can also be considered in comparison with other research coming from the USA, where there seems to be stronger evidence for a link between socio-economic conditions and adolescent mental health, controlling for individual social characteristics (e.g. Beyers, Bates et al., 2003; Dubow et al. 1997, Schneiders et al. (2003), Schonberg and Shaw, 2007; Wheaton and Clarke, 2003, Leventhal and Brookes-Gunn, 2003; Wickrama and Bryant, 2003).

A strength of the US literature is that it includes more examples of studies which examine the relationship between neighbourhood conditions and adolescent mental health prospectively, using longitudinal studies. For example, Wheaton and Clarke (2003) analysed information for 841 young people from a national sample, for whom information had been collected when they were children and when they were young adults. The survey data were combined with information on area conditions at different time points from censuses in 1970, 1980 and 1990, which enabled them to look at 'residential trajectories' of their sample. Especially for young people whose parents had a low level of education, neighbourhood deprivation in childhood was associated with behavioural psychological problems when they were older. (See Figure 5.4).

Another longitudinal study from the USA is reported by Leventhal and Brookes-Gunn (2003) who evaluated a social experiment, 'Moving To Opportunity' (MTO), conducted in 5 cities in the USA in the 1990s. Families with children living in poor neighbourhoods and occupying social housing were randomly selected to be given the opportunity to move to a less deprived neighbourhood. Leventhal and colleagues (see also Fauth, Leventhal et al., 2005, 2007) evaluated the experience of samples of families participating in this scheme in New York City, who were compared with other families who did not move home under this scheme. The experiment provided scope for a study with a longitudinal design and comparison between a 'control' group of young people who lived throughout the period studied in a very deprived residential area and an 'intervention' group of young people who moved with their families from the very poor area to a less deprived location. In the first couple of years after the scheme was implemented, boys who moved under the MTO programme were found to be less likely to reported anxious/depressive symptoms. Subsequent research (Fauth, Leventhal et al., 2005, 2008) suggests that the health advantage in terms of reduced behavioural problems were more evident for children who were younger when they moved. Among adolescents who were older when they moved, worse levels of behavioural problems were apparent for the 'intervention group' than the 'control group'. There were differences in parenting style observed in later studies of the two groups of families which may partly account for these differences, and the research seems to suggest that for the older adolescents, problem behaviours already established before they reached their teenage years were 'carried with them' as they moved. Jackson et al. (2009)

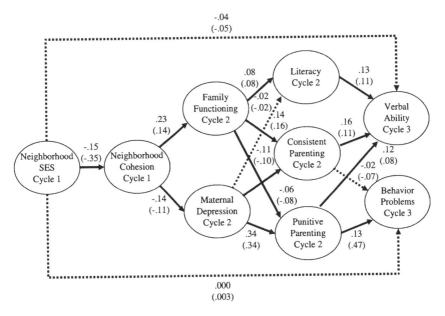

Figure 5.5   Results of structural equation modelling to explore the links between area deprivation, social cohesion and mother's and child's psychological health

Note: Estimated model, adjusted for family disadvantage. Nonsignificant paths are indicated by dashed lines; all other paths significant at p<0.5. Standardised coefficients are in parentheses.

Source: Kohen, Dahinten, Leventhal et al. (2008: 163). Reprinted with permission of John Wiley & Sons Ltd.

reviewed a large number of papers on the outcomes of the MTO scheme for adolescents and also emphasise the variation of outcomes by age and gender.

A further example of research based on national longitudinal survey data comes from Canada and is reported by Kohen Dahinten, Leventhal et al. (2008). This considers the pathways of statistical association between neighbourhood socioeconomic status (based on census variables) and social cohesion (based on survey respondents' perceptions), family characteristics and attributes of the mother and child, including their psychological health. The relationships between these variables are summarised in the analytical diagram presented by Kohen and colleagues and reproduced here in Figure 5.5. This suggests complex neighbourhood and family processes by which neighbourhood disadvantage (in terms of material poverty as well as social fragmentation) is related to family processes and maternal mental health, which in turn relate to parenting practices and to children's psychosocial development. It is also important to note that the relationships shown in this diagram were found after adjustment for family socio-economic position. The authors comment that although neighbourhood socio-economic conditions were not independently associated with children's psychosocial health, it is apparent from this diagram how neighbourhood, family and individual factors interact together in ways that are associated with mental health outcomes.

Considering overall findings from studies of younger people and adults, the studies discussed here generally confirm that differences in mental health associated with social and economic characteristics seem to emerge quite early in life, although the relationships are not always consistent and there may be quite complicated interactions between the individual and conditions in their social group and the areas where they live. These studies also illustrate typically recurring findings from research in this field, which can be summarised as follows. Different aspects of social disadvantage at the area level (e.g. social fragmentation and material deprivation) may have varying significance for mental health. These relationships may also vary depending on the type of mental health measure used, the population group studied and the setting considered. Although people in more deprived areas often report worse mental health, once the individual's personal characteristics are taken into account, the relationship with area socio-economic conditions is attenuated and seems much less significant. Using statistical modeling, it is therefore hard to discern an 'independent' influence of neighbourhood social conditions on residents' health. A further limitation of this kind of statistical study is that there is little scope to understand the lived experience of area disadvantage.

However, in some studies, aspects of collective disadvantage, including social fragmentation or lack of social cohesion, do seem to be associated with the mental health of local residents. These associations could be interpreted as supporting the argument that social relations, including ideas of 'anomie', are important for individuals' mental wellbeing today, as they were when Durkheim was writing in the 19th century, and that social relations may be bound up in complex ways with material living conditions discussed in Chapter 3. The recent research considered above and further studies reviewed below, have expanded our knowledge about the explanations for these associations between social environment and mental health and moved forward our understanding since Durkheim and Faris and Dunham were writing.

'Broken windows', social disorganisation and distress

Other, empirical research has also advanced arguments concerning the social interpretation and construction of relative differences in material living conditions, the connections between social disorganisation and geographies of insecurity, and processes of selective migration and variation in vulnerability associated with mental state. A number of empirical studies have investigated the 'broken windows' hypothesis introduced above, as a likely explanation for some of the association between material deprivation and poor mental health. In most cases, these studies have involved linking ecological information on social disorder and physical damage in neighbourhoods to survey data on individuals' perceptions of their neighbourhood, sense of insecurity and psychological distress. Some studies relate individual's perceptions of social disorder in their neighbourhood to their self reported mental health. An example is reported for a sample of adolescents in the US by Aneschenssel and Sucoff (1996), who showed that, on average, those who lived in poorer areas reported worse levels of social disorder in their neighbourhood and also had worse mental health. Mair et al. (2008) reviewed a number of studies that had related aspects of neighbourhood social disorder perceived by survey respondents to symptoms of depression, most of which showed a positive association. However, as discussed below, studies which use 'same source' information on area conditions and mental health are difficult to interpret in terms of the direction of causality. More rigorous studies of factors associated with self report of mental health problems use independent sources of information on area conditions.

Sampson and Raudenbush (1999, 2004), for example, report on studies in Chicago which have used methods such as videotaping and scoring of conditions in urban streets to collect observational data on physical aspects of neighbourhood degeneration and related these to measures of crime, reported collective efficacy of communities and perception of, and concern about crime. They discuss whether direct observation of signs of disorder be used as a proxy measure of 'neighbourhood efficacy' – the willingness of local residents to intervene for the 'common good' – or of perception of local social conditions by residents (Sampson and Raudenbush, 1999: 612). They developed a number of visual indicators of physical disorder which were recorded by trained coders. These included noxious litter in the street, such as discarded cigarette ends, used condoms and syringes, or beer bottles, abandoned cars and graffiti of various kinds. They also recorded what were interpreted as signs of social disorder such as people loitering in threatening groups in the street, fighting, people seen drinking or intoxicated in the street, drug dealing and prostitution, as well as crime statistics. These indicators were compared with survey data on perceived social disorder, showing that there was some association between perceived and observed conditions in areas. Other factors such as local economic conditions and racial composition of neighbourhoods also showed an association with perceptions of disorder. Thus while other factors such as economic poverty are probably important for perception of areas, physical degeneration may also play a role in determining perceptions of low levels of efficacy and insecurity. This may also relate to higher levels of distress in poor, predominately black urban settings, and may contribute to racial differences in mental health. Haney (2007) also reports results from a multi-city study in the USA suggesting that the impact of signs of disorder and disintegration of the fabric of neighbourhoods is associated with worse psychological outcomes, and that this partly accounts for the association between economic poverty and distress.

A further complication in this kind of study is that people exposed to neighbourhood dereliction and neglect may also have a number of other socio-economic disadvantages that are likely to increase their risk of depression.

Varying results may be obtained from different studies, depending on how many of these other factors are considered. Weich et al. (2002) reported a study in London, UK that used the *Built Environment Site Survey Checklist* to assess links between features of the built environment and symptoms of depression reported in a community survey. They found a statistical association between depression and high concentrations of graffiti and derelict buildings though this was not statistically important once a number of individual socio-economic factors were taken into account. Cohen, et al. (2003) showed that boarded up stores and homes, litter, and graffi, were associated with a reduced sense of community efficacy in Chicago. A study in Baltimore by Franzini et al. (2008) also suggests that visual cues denoting disorder (e.g. broken windows) were associated with perception of disorder and insecurity in the neighbourhood by residents, but they also note that this association varies in relation to the individual's characteristics, thus emphasising the significance for risks to mental health of interactions between individual and area attributes.

A similar impression is conveyed by the following quote cited in a British study by Popay, Kowarzik, Mallinson, et al. (2007: 973) which used ethnographic techniques to explore the pathways by which distress is produced by the stress of living in a neighbourhood where disorderly social behavior at night causes disturbance:

> ...over the past year my health has just gone worse, I don't go out of the door...I mean it's just the area has gone so bad, at one time I was always out...I could hear them all the time and I was thinking god, what if they throw something at each other, it's going to go through my windows, ...And I'm literally sitting up all night listening to them.

A different perspective on this issue is described by Cattell (2004a: 959) who quotes a young man from East London as saying:

> ...if you live in an ugly environment sometimes you act ugly...

This research on the 'broken windows' thesis needs to be interpreted cautiously in terms of the messages for policy, since the research evidence does not support a strongly deterministic view that dealing with degeneration of the built environment would be sufficient to influence processes of social disorganisation or tackle inequalities in mental health. However, the research does seem to support an argument that perception of social disorganisation is partly related to the appearance of the physical environment. This underlines the importance of the social interpretation of physical conditions in neighbourhoods and the connections between social relations and physical attributes of local settings. There does also seem to be some case for the argument that physical signs of social disorder relate to resident's perceptions of social capital or anomie and that this perception interacts with their mental state.

Relative poverty and status anxiety: the wealth inequality debate

Wilkinson and Pickett (2009) recently extended the arguments put forward earlier by Wilkinson (1996) (discussed above), presenting evidence by which they claim to demonstrate that *relative* poverty, and specifically *income inequality* at the national level are strongly associated with ill-health, mortality and a wide range of other indicators of wellbeing of the population. They argue that more unequal societies are consistently disadvantaged, on many of these indicators, compared with societies that are more equal.

These ideas have given rise to a large body of research, some of which has assessed whether individuals' perceived socio-economic position relative to other people shows an association with their mental distress. Eibner et al. (2004) used national survey data from the USA which measured income inequality using Yitzhaki's index, a measure of perceived difference between one's income and that of a more affluent reference group. They found this measure to be associated with the risk of anxiety and depression, controlling for the individual's actual income. Ellaway et al. (2004) reported from a study in the west of Scotland that, controlling for socio-economic position, those who viewed their home as having relatively low value were more likely to be depressed and have low self esteem.

In depth studies using ethnographic techniques also seem to support the idea that people living in deprived areas make comparisons between socioeconomic conditions in their own neighbourhood and other places which they find distressing or motivate them to move elsewhere. Popay, Thomas et al. (2003) found that, within the neighbourhoods they studied, residents interpreted physical signs of dereliction in terms of stigmatising perceptions of social position and it seems from their reported comments that they were concerned about feeling that they might appear to be associated with neighbours who they felt were in a different social group to themselves.

This study interpreted accounts of neighbourhood deprivation in terms of respondents concepts of what were 'proper', socially acceptable neighbourhood conditions, which also seems to underline a relativistic and 'moral' sense of the way area conditions compare with what 'should be'. This way of conceptualising area deprivation seems to some extent consistent with ideas about relative poverty and status anxiety, but may not, however, be well captured by relatively simple measures of income inequality that are used in some statistical studies.

Such findings seem to support the idea that people make comparisons with those living around them, or in their own social group and that those who feel worse off than other people may have worse health. However this kind of work, based on cross sectional studies, is hard to interpret since it is not possible to establish whether depressed mental state causes one to compare oneself unfavourably with others or whether the unfavourable comparison contributes to worse mental state.

One might therefore argue that a clearer impression of the significance of inequality for mental health might be gained from research that combines survey data on mental health of individuals with independently measured indicators of socio-economic inequality at the level of the country or region where the individual survey respondents live. Findings from this type of research are rather variable, and are also hard to compare, especially considering that they include studies carried out in a range of different countries and at different geographical scales, using quite varied methods. Selected examples of recent studies include one by Cifuentes et al. (2008) who report on an analysis of information for more than 250,000 people surveyed by the World Health Organization from 2002 to 2003 from 65 countries at varying levels of socioeconomic development. The risk of major depressive episodes was greater for people living in more unequal countries among the most developed nations. National income inequality did not show the same association with depression among less developed nations. Steptoe et al. (2007) used data on reported symptoms of depression in a sample of over 17,000 students from 23 countries. Controlling for individual variables they found that depression was more prevalent in countries with greater national income inequality. Such findings seem to support the view that income inequality and associated psychosocial processes, including 'status anxiety', are most significant for mental health of people in richer countries.

Other studies carried out within single countries have tested whether mental health of individuals is associated with the degree of income inequality within their community. The findings are variable. Sturm et al. (2002) did not find a strong relationship between income inequality at area level and depression in their study of an American sample of about 10,000 respondents from 60 metropolitan areas. Weich et al. (2001) found interactions between individual wealth and area income inequality and mental health in England.

Other research also shows that there are complex interrelationships between socio-demographic attributes of individuals, their mental health and the degree of inequality in their community. Alesina et al. (2004) used a large international survey of self reported happiness to examine the relationship of happiness with income inequalities for European countries and US States. Europeans who were poorer or were politically aligned to 'left wing' (more socialist) political ideals were most likely to be unhappy in very unequal societies. In the USA richer people seemed more unhappy in the states with greatest inequality. The authors therefore speculate whether Europeans see inequalities as more rigid and determined by the socio-political context than Americans do. This argument that 'unhappiness' is distributed according to social class divisions in some national cultures more than in others is interesting. However, as noted in Chapter 1, unhappiness and mental illness may denote different aspects of mental state and they may not relate to area conditions in the same way. This may partly explain why Kahn et al. (2000) produced contrasting results suggesting that it was the poorest women whose mental health seemed most sensitive to income inequality in US States, with those in the most unequal states more likely to report depressive symptoms. Zimmerman et al. (2006) found that area income inequality was associated significantly with self reported general health but not with mental health, and results varied for White and Latino respondents, reinforcing the impression that the relevance of inequality for health depends on the measure used and is socially differentiated. There is evidence that income inequality may be associated with mental health throughout the lifecourse; Muramatsu (2003), found that among older people over 70 years, income inequality at country level was associated with depression in the USA. There may also be interactions between area inequality, social disadvantage and risk of mental health problems associated with events such as disasters. For example, Ahern and Galea (2006) reported from their study of mental health following the terrorist attack on the twin towers in New York City (also discussed in Chapter 3 of this book) that depression was more pronounced among people with low incomes living in areas with greater income inequality.

Arguments about 'status anxiety' seem to underline the significance of differences in material wealth for one's psychological state, and the arguments concerning anomie, discussed above suggest that this is partly due to the sense of social dislocation and loss of a sense of self worth engendered by negative perception of one's socioeconomic standing. We therefore need to be cautious not to place excessive emphasis on income inequality as such, but to consider this as a 'marker' for a more complex set of processes which may undermine one's confidence about one's place in society and sense of self-esteem. The selected examples of research reviewed here show that these processes may be quite variable between social groups and geographical settings.

'Reverse causation' and selective migration

As already observed, many of the research studies reviewed here are 'cross sectional' in design and it is therefore quite difficult to interpret the associations between area conditions and mental health. It is therefore important to consider processes of *reverse causation* and *selective migration*, whereby people already in poor mental health may be more vulnerable to material poverty and tend to congregate in poor areas.

Statistical research involving systematically recorded information for extensive samples usually relies on indicators which are approximations to the phenomena being measured and may also be proxies for other, unmeasured characteristics, which is one reason for extreme caution when interpreting statistical associations in terms of 'cause and effect' relationships. Also, in 'cross sectional' studies collecting all the information analysed at one point in time, there is no way of knowing which conditions were temporally antecedent and might have given rise to other, subsequent conditions. (Logically it is hard to see how a condition that arises later in time could give rise to differences in earlier conditions, except possibly if respondents are asked to give retrospective reports on conditions in the past and their recall is biased by their experience of their current situation.)

Although much of the literature in social epidemiology uses terminology which refers to 'predictor' variables having an 'effect' on an 'outcome' variable, this should

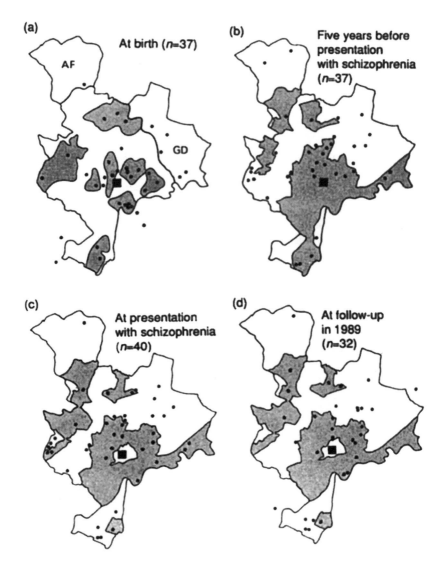

Figure 5.6   **Residential locations of a group of people with schizophrenia in Nottingham – residential locations at different time points**

Note: Residential locations at different time points, relative to deprived area of the city, shaded in grey.

Source: Dauncey et al. (1993). Reprinted with permission of the publisher.

generally be interpreted as a technical description of a statistical model, rather than an expression of actual causal relationships, especially in cross sectional studies. Research aiming to test causal processes giving rise to mental health variation is more rigorous if it is based on a longitudinal design, and can examine change in mental health, but even this method may not be able to measure all aspects of causal processes, so may give a partial picture of the chain of causation. Strictly speaking, discussion of statistical associations in terms of 'one variable *having an effect on another*' is therefore inappropriate in this kind of research.

As mentioned above, if area social conditions and health are measured using 'same source' data from a survey, then it must be considered that respondents' mental health could affect their assessment of the area social conditions rather than *vice versa*. Even when social conditions are assessed through independent observation, it is possible that the perception of these conditions by a person is mediated by their mental state. An example is reported by Fagg, Curtis, Clark et al. (2008) in a study of young people in London, UK. Data collected from adolescents in the RELACHS study (see Chapter 4) were combined with independent sources of information on area disadvantage. Young people's assessment of their local area was measured in terms of *satisfaction with neighbourhood amenities* (respondents were asked to rank aspects of their area such as facilities for young people on a scale from 'good' to 'bad) and sense of *neighbourhood alienation* (respondents were asked whether they agreed or disagreed with statements such as: 'I like this area' or 'I feel part of this area'). These indicators of perception of neighbourhood showed associations with neighbourhood disadvantage variables (for example a measure of low levels of education at the population level). Also self reported information on whether the respondent was distressed showed a significant association with area perception, even when a number of individual characteristics were included in the model. On the other hand, there was no significant association between level of distress and independently measured area conditions. The authors concluded that while it was not really possible to determine causal relationships from these results they did not seem consistent with the idea that area deprivation might *cause* distress, whereas they would be consistent with the argument that people who were more distressed were particularly 'sensitive' or 'vulnerable' to neighbourhood disadvantage. Other studies have also suggested that mental illness makes young people more vulnerable to risk in the social environment. For example, Sweeting, Young, West et al. (2006) showed that peer victimisation might be a cause of depression, especially in younger children, but the direction of causation also seemed to run the other way, so that young people who were depressed were more likely to be bullied. Both of these studies illustrate complex and 'reciprocal' causal associations between mental health and the social environment.

The 'drift hypothesis' is also an important consideration; it has been recognised for some time that people in with serious mental illnesses tend to remain in, or move towards more deprived, socially disorganised areas, especially places in urban centres where hospital care is provided (discussed in more detail in Chapter 7). For example, the frequently cited research by Giggs and Colleagues on area variation in Schizophrenia in Nottingham (Giggs and Cooper, 1986; Giggs, 1987) raised questions as to whether the concentration of schizophrenia cases in more deprived areas might be the result of migration towards these areas by people suffering from this disorder. To address this question, Dauncey, Giggs et al. (1993) examined historical records on the residential locations for the individuals diagnosed with schizophrenia. Their mapping of the residential location of these individuals (Figure 5.6) suggested that while there was some tendency for them to relocate towards the more deprived parts of the city, they also appeared to have been concentrated in these areas before, and at the point of diagnosis. Their discussion points to various possible interpretations of chains of causality: social deprivation, perhaps associated with neighbourhood poverty as well as individual circumstances, may increase the risk of complications in the prenatal perinatal period, leading to increased risk of schizophrenia in adulthood, while over the life course social processes also lead to a 'drift' towards socially deprived communities for people developing schizophrenia. DeVerteuil, Hinds, Lix et al. (2007) report findings from Manitoba, Canada which also support the idea that there is selective migration towards inner city areas on the part of people being treated for schizophrenia. There is also some evidence that processes of selective migration may apply to people suffering psychological distress, regardless of whether they receive treatment. For example, Curtis, Setia and Quesnel-Vallée (2009) found from the longitudinal *National Population Health Survey* in Canada that there was a tendency for people reporting symptoms of psychological distress to move towards more disadvantaged areas. They also note that this selective migration seemed mainly to be explained by the fact that individual characteristics associated with risk of psychological distress were also predictive of residential moves towards more disadvantaged areas, and this underlines that socio-geographical health inequalities may depend on complex interactions between processes determining both individual social position and health status.

All of these examples re-emphasise the necessity of a temporal as well as a spatial perspective in research investigating the socio-geographical dimensions of mental illness, and underline the need for caution in drawing conclusions about causal pathways that may explain geographical health inequalities.

Conclusion

The selected examples discussed in this chapter show that in several different studies, we can see evidence of associations between mental distress, social fragmentation and disorganisation, and material deprivation. It appears from several of these studies that both social and material aspects of disadvantage may have significance for mental health. The majority of research in this field has been carried out in the UK and North America, and it may not be appropriate to generalise these findings to other parts of the world where social and cultural settings may be very different.

This field of research illustrates the progressive refinement of strategies for investigating these associations, developing from purely ecological analyses to methods which combine information on individuals living within areas. There has been growing debate over whether the multi-level analyses used in more recent research are really effective in disentangling the risks of illness associated with one's individual attributes from those related to the conditions in the wider area where one lives. Oakes (2004) has argued, for example, that, because of the uneven distribution of different types of individuals among areas, multi-level models do not really distinguish differences in health associated with attributes of place from those associated with characteristics of individuals (see also Curtis and Cummins, 2007). Nevertheless, the ongoing discussion between protagonists in this debate (e.g. Oakes, 2009; Wakefield, 2009; Firebaugh, 2009) suggests that, when studying risk of morbidity or mortality, it is not sufficient to consider only individual's personal or family socio-economic attributes. Their socio-economic position is also strongly associated with the ways that they are clustered within settings which may be more or less detrimental to psychological health. We have also seen in this chapter that selective migration plays a role in the uneven distribution of mental public health across geographical spaces with different socio-economic conditions, and that one's mental state may influence one's vulnerability to challenges of social and material deprivation. Thus we see that statistical analyses of growing sophistication are required to explore the factors associated with health inequalities, including differences in mental health. A great deal can be learned as well from qualitative studies which explore these issues in depth and detail. As in earlier chapters, we can discern a trend towards interpreting research in this field in terms of complex interactions between the individual and their environment, rather than trying to conceptualise '*separate effects*' of area variables and individual attributes.

Learning objectives

This chapter presents material that will be helpful for those wishing to learn about:

- theories of 'anomie', 'social fragmentation' and 'status anxiety';
- methods of area comparison that measure the social environment in geographical areas relate these to mental health indicators;
- multi-level modelling strategies for examining the role of individual risk factors for mental illness in relation to conditions prevailing at the wider level of the local community;
- the significance of reverse causation and selective migration in explanations of geographical differences in mental health.

Introductory reading:

The following paper provides a theoretical framework that is relevant to much of the discussion in this paper.
Berkman, L.F., T. Glass, et al. (2000). From social integration to health: Durkheim in the new millennium. *Social Science & Medicine*, 51(6): 843–57.

Relatively accessible sources which would provide useful additional background to the topics covered in this chapter include:

For a geographical view of broad themes of social inequalities and human wellbeing:
Smith, D. (1977). *Human Geography: A Welfare Approach.* London: Arnold.
Smith, D. (1994). *Geography and Social Justice: Social Justice in a Changing World.* London: Wiley.

For a discussion of the issue of erosion of 'social capital' in contemporary US society:
Putnam, R. (2000). *Bowling Alone: The Collapse and Revival of American Community.* New York: Simon & Schuster.

For an overview of the evidence relating to the argument that, in wealthy societies today, inequality of wealth is more significant for health than absolute amount of wealth, and discussion of the idea of 'status anxiety':
Wilkinson, R., and Pickett, K. (2009). *The Spirit Level: Why More Equal Societies Almost Always Do Better.* London: Allen Lane.
De Botton, A. (2004). *Status Anxiety.* London: Penguin Books.

Chapter 6
Dreamscapes and Imagined Spaces: The Meaning of Place for Identity, Spirituality and Mental Health

Summary

This chapter focuses on the third 'strand' of ideas presented by 'Therapeutic Landscapes' theory, concerned with the *meanings* of places and how their symbolic character relates to mental health. Imagined spaces and perception of the significance of places are associated with different mental states. The discussion extends to some reflections on 'healthy' and 'unhealthy' ways of perceiving space. It draws on theories from environmental psychology, as well as geography, about attachment, displacement and stigma to illustrate the importance of places for one's sense of security, identity and self-worth. Certain places have special symbolic, as well as material and social importance. For example, this chapter discusses examples including: the significance of religious sites of spiritual significance, 'faith spaces' and 'home spaces'; that are important for psychological and emotional health. Most of these examples illustrate imagined spaces that are supportive for mental health or are therapeutic for those suffering mental distress, but some imagined spaces have a more sinister aspect. This chapter also discusses examples of obsessive and phobic relationships with space. The symbolic values attributed to space and place are seen to be closely linked to the material and social aspects of environment already considered in earlier chapters of this book.

Introduction: symbolic dimensions of landscape

Previous chapters have concentrated particularly on how material and social aspects of environment relate to mental health. As already noted at several points in the earlier discussion, aspects of the material and social environment are often interpreted metaphorically in ways that depend partly on their symbolic significance and meaning for people. As already discussed, it is hard to justify, either theoretically or empirically, a clear distinction between material, social and symbolic dimensions of environment. Perspectives from therapeutic landscape theory discussed at the outset suggest that we should consider all of them in order to understand how space and place relate to mental health. Therefore, while this chapter focuses particularly on 'imagined', symbolic and emotional geographies and their importance for psychological health, this does not imply a conceptual separation between these themes and ones considered earlier in this book.

The content of this chapter draws particular attention to the variability in the ways that interpretations and meanings are attributed to space and place. This variability is partly physiological, due to individual differences of sensual perception, but is also socially constructed and interpreted, depending on the social and cultural context in which the individual is situated.

'Abnormal' and 'Normal' Experiences of Space

A certain, 'normal' way of perceiving space and objects in space is associated with our understanding of a healthy mental state. Goldmanrakic (e.g. 1992, 1994) describes how this partly depends on 'working memory': "the ability to guide behavior by representations", requiring brain functions that are "engaged in holding information 'on line' and updating past and current information on a moment-to-moment basis" (Goldmanrakic, 1994: 348). It appears that such perception may become disturbed in the case of people with psychotic illness, resulting in 'hallucinatory experiences' that are often taken as symptoms of the condition. Philo (2006) highlights the point that analysis of images of places as perceived by people with mental illnesses, to gain insights into their state of mind, has a long history, and he provides an interesting discussion of the work of Laycock using this strategy. From a more contemporary perspective, Parr (1999: 678) describes how "individuals experience delusions with sight, thought, hearing and feelings, often transforming feelings of self and body", and she records examples of such experiences reported by people included in her research. Ulhaas and Mishara (2007) interpret such symptoms from a phenomenological perspective, and they reproduce extracts from various published sources describing hallucinatory experiences of space and physical aspects of the environment, as described by people with schizophrenia, either at first hand or from psychiatric case notes:

> She remembered that she could not look at the whole door. She could only look at the knob or some corner of the door. The wall was fragmented into parts. (Arieti (1962: 85), cited in Ulhaas and Mishara, 2007: 144)

> Everything I see is split up. It's like a photograph that's torn in bits and put together again. If somebody moves or speaks, everything I see disappears quickly and I have to put it together. (Chapman (1966: 29) cited in Ulhaas and Mishara, 2007: 144)

> My eyes met a chair, then a table; they were alive too, asserting their presence...I attempted to escape their hold by calling out their names. I said 'chair, jug, table, it is a chair'. But the words echoed hollowly, deprived of all meaning, it had left the object, was divorced from it, so much so that on one hand it was a living,

mocking thing, on the other, a name, robbed of sense, an envelope emptied of content. (Schehaye, M. (1970: 40–41) cited in Ulhaas and Mishara, 2007: 145)

Such experiences are evidently distressing and, literally as well as metaphorically, 'disorientating' for people with schizophrenia. Accounts like these may help us to empathise better with their perceptions and feelings. From a more detached perspective, they also draw to our attention aspects of what many of us consider 'normal' spatial perception, that we take for granted: the ability to organise the elements we perceive in space in ways that make them comprehensible and allow us to orientate ourselves physically in the world; the way we differentiate between animate and inanimate objects and understand them by naming them.

Parr (1999: 679) comments that "narratives of delusional experiences 'disrupt' the taken for grantedness of every day life [which] is a key concern for post-positivist human geography". For example, it is interesting to reflect that phenomenology raises fundamental questions about these 'taken for granted' senses of being in the world and encourages us to interrogate and try to explain them. As discussed in Chapter 1, the conceptual frameworks offered by phenomenology are also employed in geography and are important for 'relational' models of space and place. They are to some extent significant for human geography as a whole, providing us with frameworks to structure our thinking about fundamentally geographical questions of how people interact with their environment. Ulhaas and Mishara (2007) also argue that phenomenology (as well as neuro-science), therefore has a role in development of knowledge about conditions like schizophrenia that may be helpful for diagnosis and treatment. Schehaye's account cited above may give us some sense of how much we become habituated to and reliant upon certain representations of the environment and how different the world might seem if we tried to imagine a different point of view.

It is also interesting, to consider the account by Schehaye (1970) in light of ideas from Actor Network Theory that view inanimate objects as well as living things as 'actors' forming 'assemblages' that are linked together as networks. Clearly ANT does not argue that geographers should adopt a 'schizophrenic' perspective on the world by imagining that inanimate objects are 'alive' in the sense that Schehaye describes. However, it does 'disturb' more conventional views of living 'actors' (especially people) as a 'special category', treating them as though they were distinct from and independent of their surrounding environment. Questioning this view helps us to appreciate better the complex interrelationships between people and all of the other elements their environment. Similarly, phenomenological perspectives remind us that the fact that one perceives the world in particular ways that are considered as 'normal' does not necessarily mean that the world has an independent existence that exactly corresponds to our perceptions. Each individual perceives their environment idiosyncratically as well as though 'shared lenses' of biologically or socially constructed frames of reference.

Societies have always had an ambivalent attitude towards individuals whose perception of the world differs from the majority view. On one hand these perceptions may be viewed negatively and labelled as 'strange', 'deviant' or 'psychotic', treated as a sign of illness, stigmatised and marginalised or more actively punished. To quote Parr (1999: 678) people having delusional experiences are "penalised for such moments, often locked away, legislated upon, and medicated". On the other hand, societies simultaneously value, applaud, even treat as sacred, some individuals who demonstrate an ability to think exceptionally 'creatively' (i.e. differently from the average) or whose 'unworldly' views and behaviour are *revelatory*, telling us 'things we cannot possibly know in any other way' (Speed, 2003: 60; Vine, 1993). We often value these 'imaginary spaces' which may seem to invoke another, 'better' plane of existence for human beings. Popular accounts often call up stereotypes of 'tortured genius', for example, when talented artists struggle with depression. Some forms of mental 'impairments' on the autistic spectrum (identified as unusual mental ability, rather than as illness) may be associated with 'savant' powers in certain fields of knowledge such as music, art or mathematics (Treffert, 2006). For example, some accounts written by 'savant' mathematicians suggest that they perceive numbers not as digits but as "shapes, colours, texture and motions" (Tammet, 2006: 2) which they can visualise in relation to each other. This seems to allow them to appreciate mathematical relationships and process mathematical calculation very quickly. Such perception of mathematical 'spaces' is relevant here as an example of how symbolic space is associated with the workings of the mind.

In some respects these accounts are helpful, challenging popular prejudices and stigma about 'disability' associated with mental illness and impairment. On the other hand, it is interesting to reflect on why societies often seem to celebrate and revere these individuals while simultaneously 'distancing' them from 'normal' people, by emphasising the 'exceptional' nature of their human abilities in specific fields, and contrasting their special talents and abilities with their mental illness or impairment with respect to other social and physical capacities. This kind of recognition may also inadvertently contribute to perceptions of stigma and 'difference' attached to certain mental states.

In the context of the present discussion, the point at issue is that human ability to create and express 'imagined' worlds ranges widely and outstanding imaginative ability of this sort enriches human life. Examples of people whose achievements in this respect have captured public attention would include artists such as the Sylvia Plath (1932–1963), poet and author, under the pseudonym of Victoria Lucas, of *The Bell Jar.* Kim Peek (born 1951), has exceptional powers of memory and mathematical skills and is celebrated in popular culture in films like *Rain Man* (Peek, 1994). Daniel Tammet is another 'savant' who has written very engagingly about his perceptions (Tammet, 2006). Another inspirational example for those interested in the study of built forms is the work of Stephen Wiltshire from London (born 1974), who has attracted significant media attention for his ability to perceive, recall and reproduce in drawings prodigious amounts of detailed information about the built environment (Wiltshire, 1989 and see Figure 6.1). He is talented in the field of architecture, but also as a musician.

**Figure 6.1   Tokyo by Stephen Wiltshire, depicting in wonderful detail
the intricacies of the built form of the city. This is a striking
example of 'savant' ability to store and process information
and turn it into artistic masterpieces**

Source: http://www.flickr.com/photos/kymail100/74763785.

Moving from the field of the arts and sciences to consider areas of life relating to
spirituality and religion, we see that here too similar arguments apply concerning
the symbolic role of 'imagined spaces'. Many individuals accorded holy status in
religious orthodoxies (who have helped to found important sacred sites with great
reputations for healing) have been marked out by their 'revelatory' experience
in 'paranormal' 'visions' and 'dreams' with spiritual significance, or by attitudes
and behaviours which might, from another perspective seem 'fanciful', 'morbid'
or, possibly, in more popular language, 'mad'. (See the discussion below of St
Bernadette of Lourdes, and the life story of St Cuthbert and its significance for the
city of Durham).

An appreciation of the different ways of imagining 'space', associated with altered
mental states is particularly relevant for health geographers studying the psycho-
spatial experience of people with psychotic illness, but also has significance more
generally for geographical knowledge of mental health and wellbeing. Alternative
perspectives prompt us to delve more deeply into what we know from geography
and other disciplines about the ways that we all perceive socio-geographical spaces,
invest them with symbolic meaning and respond to them emotionally.

Dreamscapes as 'windows to the mind'

The symbolic character of 'dreamscapes', the imagined settings experienced in dreams or trances, may be important for the interpretation of a person's psychological state and they may also influence culturally constructed ideas about the symbolic significance of places.

This idea has been widely used in branches of psychology for some time. Examples abound in a classic reference in psychoanalysis: Freud's book *The Interpretation of Dreams*, first published in German in 1900. Below is an extract from a passage in which Freud described his recollection of one of his dreams, before proceeding to interpret its psychological significance.

> ...I make my way through a suite of handsomely appointed rooms, evidently ministerial apartments with furniture of a colour between violet and brown, and at last I come to a corridor in which a housekeeper, a fat elderly woman is seated. I try to avoid speaking to her, but she apparently thinks I have a right to pass this way, because she asks whether she shall accompany me with the lamp. I indicate with a gesture, or tell her, that she is to remain standing on the stairs, and it seems to me that I am very clever, for after all I am evading detection. Now I am downstairs, and I find a narrow, steeply rising path, which I follow...It is as though my second task were to get away from the city, just as my first was to get out of the building. I am riding a one-horse cab, and I tell the driver to take me to a railway station... (Freud 1997 edition, pp. 109–10)

In subsequent pages of his account, Freud offered an interpretation of this dream, arguing that the images of settings (dreamscapes) are symbolic references to very recent activities and experiences the previous day, and also to childhood memories and emotions and to adult desires. For example, there are many references to his actual recent experience: he had this dream while travelling on an overnight train in a first class sleeping compartment, which he had obtained with difficulty and which was inconvenient as it had no *ensuite* toilet. At the station he had seen a member of the German aristocracy boarding an earlier train and been impressed by the way this person waved aside the guard wanting to check his ticket. He also suggests symbolic significance of elements in the dream, involving both visual imagery and also aural references (it is typical of Freud's interpretation to see references in dreams to words in German which have multiple meanings and/or sound similar to other words). Freud argues that symbolic elements of the dream are references to early childhood experiences and/or repression of socially 'improper' sexual thoughts. For example (pp. 113–14), he refers to the idea that the suite of rooms invokes the idea of a woman – in this case one held in slight contempt (the translator's note points to Freud's use of the word *Frauenzimmer*, combining German words for both woman and room). Freud also recounts how he had bought a new suitcase of a brown-violet colour for the journey, the same colour as the furnishings in the rooms of the dream sequence. He further suggests that the compartment without a toilet had raised anxieties which recalled early his infant experience of bedwetting. Therefore Freud sees his dream as reflecting psychological concerns such as the stresses of the train voyage, anxiety about status and social sanctions on behaviour. The dreamscapes that Freud recalls can therefore be viewed as 'spaces' with multiple significance and meaning, though which it is possible to reflect not only on recent mundane events, but also on deeper memories and emotional concerns. Dreamscapes are examples of imagined spaces which are creations and expressions of our mental state. They also impinge on our emotions, so that we experience some dreams as 'nightmares' and others as euphoric, 'good dreams', which can affect our mood after we wake. We might consider parallels and connections between Freud's way of thinking about dreams and arguments from health geography about imagined spaces and their significance for emotions and psychological health.

Spiritual imagination and sacred spaces

Dreamscapes offer imagined worlds that confer enriching spiritual experiences. Dreams have frequently been represented as 'messages' communicated from an ethereal, 'spirit world', guiding us as to how to respond to challenges in life and follow a 'virtuous' path leading to wellbeing and a healthy relationship with the world around us, or offering solace for emotional stress or pain. Religious stories and moral tales abound with accounts of people being 'visited in a dream' by spirits, angels and other similar messengers from the 'other world'. Often the sites of these mystic encounters take on special spiritual significance, frequently associated also with a reputation for physical and psychological healing.

Some of the earliest geographical studies of therapeutic landscapes focused on such sites. Also a significant field of social and cultural geography is devoted to studies of religion and spirituality (e.g. reviewed by Kong, 1990,2001; Brace et al. 2006; Proctor, 2006a) and while some of this is rather tangential to the issues of interest here, those aspects concerning the links between sacred spaces, emotions and mental state (e.g. reviewed by Buttimer, 2006) do contribute to the arguments in this chapter. Thus there is an intersection between geographies of sacred spaces and geographies of health, which is especially important in research on mental health.

An example from the literature on therapeutic landscapes that illustrates this connection with geographies of sacred spaces and also emphases the significance of dreams, is Gesler's (1996) account of Lourdes, France. The site has gained an international reputation as a place of healing. Gesler records the story of Bernadette Soubirous who experienced repeated dreams of the Virgin Mary calling the people to prayer and repentance and signifying the special properties of a certain spring, hidden in a grotto. For members of the Catholic faith, the site began to acquire a symbolic and religious significance as a sacred place, partly because of these visionary manifestations of divinity, personified by the Virgin Mary, a very popular figure in Catholic hagiology, and also because of the reputed healing properties of the spring water. Today, Lourdes has become internationally recognised as a place which is attributed special religious significance for Christian worshippers who revere the spiritual and healing properties of the shrine located in the town. A magnificent cathedral marks this religious site.

Another illustration, rooted in a different sacred tradition, is Gesler's (1993) account of Epidaurus in Greece. This discusses the interplay between features of the built and natural landscape, the association of the place with ancient Greek deities over a long period in Greek mythology and its particular dedication to the god Asklepios, credited with protecting the local population from the ravages of a plague epidemic 430–427BC, and continuing to be represented as a place with special healing properties into the Christian Era. Gesler's account is particularly relevant to the present discussion for the way he describes the emergence of the reputation of Epidaurus through a long tradition of mythical and religious associations, reinforced by the cultural significance of buildings and landscape of the site, and also because of his reference to writing by Scully (1969) concerning the association in ancient Greek culture of certain gods to particular holy places. Those seeking healing came to Epidaurus in the belief that they would be cured in dreams. Gesler references Oberhelman (1981) on the belief in dreams as prescriptions to cure illness. Oberhelman (1981: 419) cites Artemidoros, an ancient Greek author, writing as follows:

> Someone with a stomach ailment prayed to Asklepios for a medical prescription. He dreamt that he entered the temple of Asklepios and the god stretched forth his hand and offered his fingers to the man to eat. The man ate five dates and was cured; indeed the fruits of the date palm tree, when they are in excellent condition are called 'fingers'.

For the present discussion, it is significant that the dreamer imagined himself *entering the temple*, which seems to emphasise the symbolic association between occupation of an imagined space and proximity to healing powers. Oberhelman also discusses the references that Sigmund Freud made to the ways that ancient Greeks interpreted dreams, emphasising a direct connection between these ancient traditions and the more contemporary roots of psychoanalysis.

Another Greek site, famous in antiquity, and widely studied and much visited right up to the present day is Delphi, where there was an ancient Greek temple to the god Apollo. Figure 6.2 shows that the site exhibits features found at many Greek holy places, being located in a remote area on commanding topography which at the same time is a sheltered 'sun-trap' with a benign microclimate and uplifting views across the landscape. (We might consider this in light of theories about biophilia, especially Appleton's (1996) 'prospect and refuge' theory discussed in Chapter 2.) Its reputation as the source of cultural traditions drawing on classical themes from ancient Greece also makes the place significant for ideas about topophilia and 'virtuous' architectural forms, introduced earlier.

Figure 6.2   Delphi in Greece

Note: This site has long had special symbolic and historical significance; originally in Ancient Greek society, and globally today. The picture shows the beautiful and commanding landscape of the site and remnants of the original architecture. The place symbolises especially the social and cultural significance of the Delphic oracle, an example of the importance people attach to the ability to use altered visions of the world imagined in 'dreams' to analyse and determine human action. One could also interpret the landscape features of Delphi in terms of theories of biophilia and topophilia, discussed in Chapter 2.

Delphi is also relevant to the present discussion of dreams and their associations with sacred places. The sanctuary at Delphi included the *oracle* in which the *Pythia*, a priestess, was reputed to inhale vapour emanating from the ground, enter a trancelike state and utter words which were interpreted as prophecies and advice by her listeners. The oracle at Delphi therefore seems to have had a spiritual and therapeutic role as an important source of help and counselling for those faced with major life events or difficult decisions. It is credited with influencing many eminent leaders in Mediterranean cultures of the period.

The site holds continued fascination for scholars in science as well as the humanities. Recent research has investigated whether there might be geological evidence that seismic activity in the area would have resulted in the emission of gases which might have produced the intoxicated trance of the *Pythia* (e.g. Piccardi et al., 2008). This would provide another interesting example of how physical attributes of a place may play a role in the establishment of religious and cultural practices, which in turn influence human perceptions and social behaviour.

Delphi has also retained a special significance in classical traditions of thought in western cultures which has been transferred into nomenclatures used in contemporary research in social science. For example, today, the name 'Delphi' is applied to methods of inductive processes of formulating new ideas through consultation processes. The method is used widely in different fields of social science and health research, and it is interesting here to note that these techniques have been used in holistic and traditional medicine to draw out consensus from practitioners about good practice. For example, Flower et al. (2007) argue that Delphi techniques are a suitable method for review of practice in Chinese medicine, which is not very amenable to biomedical research methods. Estby et al. (1994: 402) reported a Delphi survey of holistic nurse practitioners which reached conclusions which support arguments in this chapter about the role of spirituality in healing:

> In addition to affirming principles related to unity, interdependence, evolution, energy fields, and interactions, the...expert nurse practitioners strongly emphasised spirituality.

Many non European societies also have traditions in which dreams play a central role in the associations between sacred landscapes, spirituality and psychological wellbeing and have provided a way of understanding human relationships with the natural world. Bruce Chatwin's (1988) account of the *Songlines* records traditions in aboriginal Australasian creation stories about the 'dreamtime'. We also consider in more detail below the importance of these beliefs about the sacred links between land and spirit for ethnic identity of aboriginal peoples.

The symbolic importance of sacred places are not limited to their association with *revelatory* experiences and 'dreamscapes, but are also identified by iconic physical features which encourage religious *reflection* (Speed, 2003). This reinforces and celebrates the tenets of the religious faith to which they are dedicated. It is most apparent in the structure and adornments of all churches, temples and

Figure 6.3   The Solidarity Memorial in Gdansk, Poland

Note: The Solidarity Memorial is a memorial to the Polish shipyard workers who played a pivotal role in the overthrow of the communist regime in Poland in the 1980s. The memorial uses Christian iconography in the form of the holy cross. Churches had also been sites of resistance to communist domination.

other religious architecture. As Holloway (2006) has pointed out, a key aspect of sacred and spiritual settings is their emotive significance, and it is this emotional importance, rather than their doctrinaire function, which is of particular relevance for their association with mental health.

Part of the psychological value of sacred places is the way that they symbolise a wider community as an imagined, faith based 'social space', within which the individual believer can feel that they belong and feel accepted. As Kong (2001) has pointed out, religions may also have the effect of reinforcing social divisions between those who are 'of the faith' and can appreciate the symbolism, and 'outsiders' who do not share this frame of reference and are psychologically, if not physically excluded. However, for those who do have an affinity with a particular faith, sacred places can offer a self-affirming and supportive sense of social bonding with other believers, and this may be beneficial psychologically. Proctor (2006b) also emphasises the links between religion and trust from a geographical

Figure 6.4   A church building in Katowice, Poland

Note: The church was photographed in 1987, a period of intense nationalist political activity in Poland. Built by private subscription, the building defied the norms of Soviet Baroque building styles of the period with a modernist structure recalling Noah's Ark, from the biblical story of the preservation of virtuous believers from the devastating flood. Metaphorical references seem to include ideas about religious faith as a way to preserve national identity.

perspective. It was noted in Chapter 4, for example, that in an African American Community in North Carolina, the church was recognised as a symbolic site for generation and maintenance of social capital which could be beneficial for health (Ornelas et al., 2009). Smith (2008) also discusses the importance of a Buddhist temple for Japanese Americans in Los Angeles.

Religious symbols are widely used to express national identity, so that there is a link between religious symbolism of sacred places and ideas of national and political identity, as discussed below. The pictures in Figures 6.3 and 6.4 show examples from Poland of built structures with religious iconography associated with Catholicism which also had special significance for the Polish nationalist movement in the 1980s, while the country was still dominated by the Soviet communist regime. Religious and political iconography are blended in ways apparently designed to support a spiritually 'uplifting' sense of national identity linked to religious faith.

Figure 6.5   Durham Cathedral, Durham, England

Note: Durham Cathedral, is an outstanding example of a sacred site dedicated to St Cuthbert and built around his reputation as a holy recluse and healer. The Cathedral is located on a prominent site, high on a peninsula formed from a deep incision by the River Wear and surrounded by woodland.

Sacred places are also valued as spaces for private contemplation and reflection on reassuring and comforting religious or spiritual beliefs and they offer opportunities for retreat and refuge from the stresses of ordinary life. Gesler (1996) describes this as part of the experience of pilgrims to Lourdes, for example. Certain religious sites have built significant reputations as sacred places that seem partly grounded in an appeal to the reclusive, contemplative aspect of human nature. One example is Durham Cathedral dedicated to St Cuthbert.

Durham Cathedral (Figure 6.5) is dedicated to St Cuthbert (c.630–687). He was evidently an exceptional and inspirational person, very influential in the Christian community of northern England of the time, and several healing miracles have been attributed to him by the Venerable Bede (721). He is reputed to have had a strong attraction to wild, natural environments, an extreme unwillingness to engage in the more 'worldly' aspects of life and a strong desire for extreme seclusion and withdrawal from human contact. Although for a period he held a prominent role as a Bishop in the Christian church, he later retreated to a solitary hermitage on the lonely Island of Farne, where he seems to have shunned human company and practised a degree of asceticism that hastened his death. It is recorded that later his tomb was reopened and his body was found to be uncorrupted, which was taken as a sign of his exceptional spiritual purity and sanctity. Risks of foreign invasion and military conflict at this period threatened the security of religious institutions and his followers were forced to transport his coffin around the north of England until, according to legend, one of his followers was instructed in a dream to take his remains to Dunelm, the current site of Durham Cathedral. He was venerated as the protector Saint of northern England and his shrine at Durham cathedral grew in significance as a site of pilgrimage associated with healing miracles. The town gained influence as a centre of ecclesiastical power, and a seat of learning and regional government especially during the 8th and 12th centuries. Castle fortifications were added, contributing further to the imposing character of the place and Durham University has a long history dating back to the influence of the Venerable Bede, an internationally recognised scholar of the period, whose writings included 'biographies' of St Cuthbert. The Cathedral is located on a prominent site, high on a peninsula formed from a deep incision by the River Wear and surrounded by wooded green space (Figure 6.5). The religious and historic significance of the site, combined with its spectacular architecture and landscape has given it an international reputation which has resulted in its present status as a world heritage site.

The story of St Cuthbert still influences the development of the built form of the city today. Figure 6.6 shows a sculpture recently set up in the commercial centre of the town that depicts St Cuthbert's faithful followers carrying his coffin towards the cathedral. It seems to express not only the story of the saint's life and influence, but also the symbolic significance of a journey as a metaphor for the search for spiritual objectives (see the discussion later in this chapter).

This story illustrates some of the themes in this chapter. To understand St Cuthbert's significance to faithful Christians in the early English period, we need to consider socially and historically variable interpretations of human states of mind and personality traits and the credibility of miracles. For 7th and 8th century Christians his solitary and ascetic behaviour was viewed as a sign of holiness, to be venerated. In today's more secular British society it might be considered by many people to be eccentric, perhaps morbid, although we should also consider that the appeal of retreat and spiritual contemplation is still strong for many others. A more scientific perspective in modern society probably makes many people more sceptical as to whether stories of miracles associated with St Cuthbert are accounts of real events as opposed to allegorical tales. We see how the history of St Cuthbert and the posthumous cult surrounding him led to the establishment of a sacred site which became an important symbol for the Church of England and a destination for pilgrimages (see the discussion of pilgrimages later in this chapter). Thus a number of elements merge together to create a sacred place; religious and spiritual imagery and discourses combine with landscape elements which according to theories reviewed here and in Chapter 2 are psychologically therapeutic. Furthermore, the links between St Cuthbert's life story and the historic

Figure 6.6 **Sculpture in Durham city representing St Cuthbert's coffin being brought to Dunelm, the site of Durham Cathedral**

role of Durham, as a site of pilgrimage and healing, reflect processes of symbolism and meaning of place, historical narrative and faith, and the transformation of personal reputation into reputation of a place, consistent with Gesler's (1996) intepretation of Lourdes.

These examples illustrate some general points made also by authors such as Speed (2003: 62) about the sacred significance of places residing in human 'emotional investment in things and places that forms the basis for their basis for their moral significance'. We therefore see a strong connection between geographies of sacred spaces, 'emotional geographies' and spaces of significance for cultural identity, discussed below. For example the Acropolis in Athens, has been interpreted by Loukaki (1997) who considers the *genius loci* (spirit of the place). One interpretation he considers is that the site is symbolic of both specific cultural traditions and identity for Greeks, and also of rather universal appreciation of ancient Greek culture for many societies influenced by Greek philosophy and democratic political organisation. The Acropolis is also a site of contention, since the priorities and perceptions of different interest groups vary, in terms of access to and use of the site and how it should be maintained and preserved and whose interests should be considered.

'Rocks of emotion': symbolism of natural landscape features related to human emotions

Most of the examples of sacred sites considered so far are built environments, but natural landscape elements are also invested with spiritual symbolism. Buttimer's (2006: 198) overview of geographies of religion also draws attention to "the sensory interactions between human bodies and natural elements of fire, air earth and water...seen as essential in the quest for wholeness...and health that lies at the heart of most world religions."

Various studies have explored emotional responses to natural landscapes, emphasising different symbolic representations. They show, for example, how the imagined significance of nature may be constructed in terms of sources of wellbeing and strength and solace, memory and expression and preservation of personal and cultural identity, and moral ideas about redemption or retribution.

Thus the spiritual symbolism of natural landscape can be important for one's sense of psychological wellbeing and it is included, for example, in settings used to practice yoga. Lea (2008) describes observations at a yoga retreat in Spain and her conversations with the retreat leader who emphasised the idea of emotional refreshment and renewal gained from being in the beautiful and rather austere natural setting in the mountains. Her account also emphasises the idea from yogic teaching of 'drawing natural energy' from the earth. Hoyez (2007) has explicitly interpreted spaces dedicated to Yoga as examples of therapeutic landscapes and has particularly emphasised the way that certain symbolic elements have been transferred to different sites around the world, indicating the globalisation of yoga practice and imagery. The design and location of Yoga centres are typified by features of the landscape or the design and decoration of the buildings that incorporate river and mountain scenery, symbolic of ritual cleansing in the River Ganges, and the Himalayan source of the Ganges. Hoyez (2007: 116) comments that "These elements recall, on the one hand, the divine nature of the world's construction invoked by most religions, and, on the other hand, are associated with an idea of well-being consistent with yoga practice".

In Chapter 2 it was noted that the therapeutic psychological appeal of water in the landscape is closely bound up with its symbolic association with cleansing and purity. Here we consider the culturally constructed spiritual significance attributed to rock as a landscape element. Repeatedly, and in many different cultures, the strength and durability of rock is used metaphorically for psychological and spiritual power and constancy.

Rock has carried symbolic significance associated with foundation and preservation of religious, cultural and national identity in many times and cultures. Christian religious symbolism includes the naming of St Peter as the 'rock' on which the Christian church was founded. The biblical 'ten commandments' dictating how to live a virtuous life were inscribed on tablets of stone. Earlier religions and philosophies have also invoked the psychological significance of rocks. For example, the standing stones at Stonehenge and at Avebury in southern

Figure 6.7   Standing stones at Avebury in the south of England

Source: http://www.geograph.org.uk/photo/442932 © Simon Barnes. Reproduced with permssion.

England (Figure 6.7) were sacred to the ancient British Druid religion and are still used to celebrate druid rituals today. In Australia, Uluru (Figure 6.8) is another example of rocky topography associated with cultural identity, which holds symbolic and sacred significance for aboriginal people. Nogue, and Vicente (2004) discuss the example of mountains in Spain that hold symbolic significance for Catalan identity.

Consideration was given in Chapter 2 to Feng-Shui traditions which specify certain dispositions of elements in the natural landscape and their importance for wellbeing. Here it is also relevant to discuss the significance of Daoist and Confucian philosophies for the symbolism in classical Chinese garden aesthetics. For example, Hu Dongchu (1991) explains that Daoist thinking emphasises the value of simple, natural elements in nature. The aim of a Chinese garden was to contrive an ideally composed and carefully ordered space which still retained a simple, 'natural' appearance. The Daoist perspective was complemented by Confucian philosophy. In traditional Chinese culture, these schools of thought offered a moral education and a guide to psychological and physical wellbeing which filled the place occupied by religion in other cultures. Fundamental to traditional Chinese notions of morality and wellbeing is the idea of natural forces of *yin* and *yang*. These represent opposing characteristics such as hardness and softness, dynamic motion and calm passivity. Health is understood as balance between these opposing and interdependent forces and illness is interpreted as

Figure 6.8   Uluru, Northern Territory, Australia

Source: Photograph by Ruth Jenkins.

imbalance. These ideas permeate many areas of Chinese culture. Traditional Chinese medicine is based on this conceptual model and offers therapies aiming to help the patient to maintain or restore a 'healthy balance'. According to Dongchu (1991), these ideas of *Yin* and *Yang* were also translated by philosophers such as Lao Zi and Zhang Daoling into aesthetic principles incorporated into art and also into garden design. *Yin* elements included soft waters and flowering plants. Rocks represented hard, strong, *Yang*. *Yang* aspects of human temperament, such as strength of purpose, persistence and fearlessness were also especially valued in Chinese society, so rocks can be seen as symbolically representing these psychological attributes. Chinese gardens constructed on these principles also included the idea of space and calm to encourage repose and reflection. A further interesting aspect of these gardens was that small features were thought to symbolically represent larger landscape elements, so a single stone might be taken to represent a mountain. The traditions of Chinese garden design have been influential throughout oriental countries (For example, see Figure 6.9 for examples in Japan) and they have also been introduced into western cultures, so there has been a process of globalisation of the symbolism which parallels that identified by Hoyez (2007) in respect of spaces dedicated to yoga. These design principles (as well as elements of *Feng Shui*) are often used, for example, in gardens attached to hospitals and other settings which have a therapeutic function. By including these elements, symbolically healing, therapeutic spaces are created, intended to benefit psychological health and wellbeing.

Figure 6.9　Traditional gardens in Kyoto, Japan

Note: The aesthetic involves careful placement of rocks, plants and water to create a harmonious whole. The design is influenced by the principles of virtuous design.

Stone often has psychological and emotional significance for preservation of personal and cultural memories; because of its durability it is often used in monuments and memorials. Speed (2003, 559) considers Lutyens' stone of remembrance, created at the behest of the Commonwealth War Graves Commission "to give a feeling of solace, not depression". The form of the sculpture recalls a sacrificial altar. The image is spiritual but non-ecumenical, designed to appeal to humanity in general rather than a single religious faith. Speed comments that "Sacrifice is an emotive concept" which may induce "feelings of pity, either because we feel sorrow for innocent lives that have been taken unjustly, or because...what we fear ourselves provokes pity when it happens to others". Speed also comments that the memorial symbolises durability and a timeless quality reflected in the ancient geological origin of the stone.

These symbolic readings of natural elements in landscape, such as rock, all illustrate culturally constructed frames of reference, formulated through philosophical or moral discourses. These seem different from the universal, genetic predisposition towards certain landscapes implied by theories such as biophilia, discussed in Chapter 2. They may be more consistent with ideas from topophilia which draw on cultural geographical explanations for the emotional bond between people and places. However, proponents of theories about universal and instinctive responses to landscape might argue that these symbolic schema used to interpret landscape in different cultures have their origins in more fundamental, shared human responses to natural settings. Thus a complex combination of instinctive response and culturally constructed symbolic interpretations may be operating to influence the meanings that landscape holds for one's psychological health. To understand these relationships requires a relatively elaborate conceptual model.

Culturally diverse concepts of wellbeing and natural landscapes

Furthermore different cultures vary in their understanding of how psychological and physical health relate to natural environments. Some cultural traditions do not separate out ecological, spiritual and psychological aspects of wellbeing to the same degree as in modern western societies. This is illustrated in research which has explored the emotional and psychological relationship to the land among aboriginal peoples in Australia and North America. Several studies have highlighted the importance of the land for wellbeing of aboriginal people. This research is relevant here for its focus on the functional, emotional and psychological relationship to land as well as the political issues involved. Richmond et al. (2005, 2007) used a political ecology framework to structure their study of the response of the Namgis first nation peoples to the environmental damage caused by the aquaculture industry on the Coast of British Columbia. They use this example to show how access to natural resources is essential for the general health of the population and how restriction of access undermines the whole social and economic fabric of their community. Wilson (2003) has also analysed the significance of landscape for first nation peoples in Canada using a therapeutic landscapes perspective. Considering the case of Maori populations in New Zealand, Panelli and Tipa (2007: 445) have argued that "...a geographical approach to well-being enables the linking of culture and environment...into both ecosystems and human health". They suggest that western measures of psychological wellbeing are inappropriate for Maoris, being based on individualist ideas about the self as autonomous. In contrast, Maori health and wellbeing is better understood in terms of ideas that combine spiritual, social, cultural and biophysical elements; "...ecosystems do not simply connect biophysical components and processes, but rather involve a combination of physical, spiritual, and socio-cultural phenomena that fundamentally blur any modern Western scientific division between human and nonhuman elements" (Panelli and Tipa, 2007: 451). Health is viewed as a house with four 'cornerstones' comprising mental wellbeing, spiritual wellbeing, family well being and physical wellbeing. For Maori, alienation from the land has spiritual and social as well as economic and political ramifications, and is damaging for psychological as well as for physical health and wellbeing. Kearney and Bradley (2009) report how for the Yanyuwa people of Australia the place called *Manankurra* in their tribal homelands remains an important part of their cultural identity and lends a sense of psychological resilience, even though they are physically exiled from the place. Richmond and Ross (2009) make similar points about the need for public health models to understand the cultural, psychological and spiritual, as well as economic significance of land for First Nations in Canada. These examples therefore show how the spiritual significance and culturally constructed understandings of place have an important bearing on mental as well as physical wellbeing for these groups. Kingsley et al. (2009) make similar points in respect of Australian Aboriginal peoples.

Symbolic readings of the natural environment in terms of ideas of redemption have been identified. For example, Bell and Evans (1997) refer to this in their interpretation of plans for a forest park in England, which was discussed in Chapter 2. These ideas are also reflected in literature, as illustrated, for example by Tonnellier and Curtis (2005) in their interpretation of *The Country Doctor* by Honoré de Balzac. Not only does the book reflect a conceptual model of what a therapeutic landscape might be, but it also uses this landscape symbolically as an allegory for spiritual progress and redemption.

In contrast, aspects of natural environments symbolic of retribution also abound in various cultures. Moral fables in religious and secular imagination often involve environmental disasters visited upon the iniquitous to punish them for their wrong doing. Obvious examples from Judeo-Christian traditions include biblical accounts of natural disasters such as the flood described in the book of *Genesis* and, in *Exodus*, the storms and plagues of locusts and rats afflicting the Egyptians who had enslaved the Israelites.

Distress over loss of culturally important landscapes or feelings of guilt and shame over destruction of environments is an issue of growing concern in many societies. According to Beck's (1999) arguments, discussed elsewhere in this book, present day responses to environmental change such as those arising from climate change or environmental contamination include a growing sense of anxiety associated with the impacts of climate change. Albrecht et al. (2007: s95) have proposed the idea of *Solastalgia*, which they define as "the distress that is produced by environmental change impacting on people while they are directly connected to [i.e. while they are living in] their home environment", and which expresses the sense of loss of 'solace' offered by the home environment when it is contaminated or damaged. The same group of researchers has also developed an Environmental Distress Scale, to measure the "bio-psychosocial cost of ecosystem disturbance" (Higginbotham et al., 2006: 245). Trudgill (2008) also discusses emotional responses to the change in British flora resulting from climate change and argues that this is associated with socially constructed ideas of what the 'proper' natural flora of the country should be like. The symbolic links between landscape and psychological wellbeing and security are thus seen to be issues of widespread significance internationally and ecological models of mental health are increasingly understood to be relevant in many social settings.

Home and identity

More generally, the discussion above has explored how memory and the imagination play a key role in the association of places with cultural or individual senses of identity. 'Heritage sites', memorials, and museums all have special psychological and emotional significance for those to whom they signify the idea of the 'home' of one's family or cultural group. These associations help to explain the emotional bond with certain places invoked by theories such as topophilia, introduced in Chapter 2. As well as public spaces, more private settings, such as the domestic space of 'the home' are important in this respect.

There is a large geographical literature on questions of identity, place and symbolic space and the ways that these are linked to ideas of 'home' and home space. Other sources, providing a more comprehensive review than is possible here, include Moore and Whelan (2007); (Mitchell, 2004, 2006, 2007); Brace et al. (2006); Blunt and Dowling (2006). These reviews develop more comprehensively the ideas touched on in this chapter, showing that 'home' is made more complex by processes of multiculturalism, cosmopolitanism, and exceptionalism. They often emphasise strong associations between spiritual and religious feelings of 'belonging' and the formation of personal, social and political dimensions of identity, which have also been illustrated in the discussion above.

Quite often rather banal, everyday objects and practices in the home have emotional significance at a personal level. For example Rose (2004) analyses responses to family photographs which evoke a sense of physical as well as emotional 'togetherness'. Dyck and Dossa (2007) consider how women in immigrant communities in Canada work to create healthy spaces in the home through practices of food preparation and consumption as well as social and religious practices. This emphasises a link between affirmation of home as the social and symbolic space for nurturing cultural and family ties and ideas about wellbeing and health. Morley (2000) also emphasises the domestic home space as place where a sense of comforting security is derived from observance of certain family rules and obligations that help to bind family members together. Hockey et al. (2005) discuss the role of the home and its contents for older people as an 'environment of memory' and its relationship to emotions such as grief and loss of loved ones.

Other authors have also noted that for some household members and for women or children in particular, the home space may hold varying and ambivalent significance. In many cultures the family home may be socially constructed a place which contributes to a woman's sense of self-esteem as well as emotional wellbeing, where her family roles are especially appreciated and respected, and where she is able to exercise certain rights and autonomy. However, this understanding of the home may also reinforce gendered differences in the social roles that are permitted for men and women and the exclusion of women from certain public spaces. Such gendered differences in the social significance of the home are often observed for traditional cultures (e.g. Bray, 2005 in respect of traditional Chinese homes). However, similar arguments might also be applied to western cultures where home work remains a predominantly female activity in many families and roles in the work place and other public spaces are often still male dominated. It is also important that for some women, the home may be a place where they may experience subordination, psychological and economic oppression or violence (e.g. Nicolson and Wilson, 2004; Fawole, 2008).

For children similar variability is evident in the way that home is imagined and the social relationships and processes in the home space that are important for wellbeing (e.g. Holloway and Valentine, 2000; Morrow, 2004). For instance, Christensen et al. (2000) reported that young people identify the home strongly as a 'family setting', and most liked to spend time at home, but at the same time the home was seen as place where their activities are regulated and monitored by their parents. Aitken (2000) draws particular attention to the psychological tensions this produces for children and parents in the home, and Skelton (2000) reviews research which shows how the social construction of children's roles and restrictions on their behaviour in and around their home varies for girls and boys. All of these authors also emphasise the importance for children, especially in adolescence, of spaces outside the home where their activities are not monitored or regulated by adults to the same extent and they can build friendship networks with their peers that are also important for their psychological wellbeing.

'Home' can therefore be thought of as a complex series of public and private settings that are constructed as symbolic spaces, representing aspects of family ties and social identity and having associations with psychological and emotional health that are variable and ambivalent. 'Home' spaces may be both beneficial and detrimental to wellbeing.

Emotional attachment to place and the significance of displacement, journeying, discovery

Our understanding of the significance of places for ideas of identity and self awareness are further enriched by research which focuses on *displacement*. Displacement may occur through involuntary processes such as destruction of one's home place, or forced migration, but also may be voluntary, in the case of journeys and migrations that are undertaken by choice.

Travelling is often viewed metaphorically as a process of personal discovery and development. Pilgrimage, in which the destination is a sacred site, is one expression of this process, which is recognised in many different cultures and religions. We have reviewed examples above of sacred sites which have developed a reputation as destinations for pilgrims seeking redemption, spiritual purification and renewal and physical and mental healing. Pilgrims' experiences of their journey, as well as at the destination shrine, are often seen as having psychological significance. Gesler (1998b) draws attention to the levelling of social differences and the opportunities for social interaction that are created when groups of people from different walks of life take place in a pilgrimage. Accounts of pilgrimages going back to Chaucer's *Canterbury Tales* also conjure up this sense of interactions between people in very different social strata which would be unusual in the course of 'everyday' life at home. Travails in the 'slough of despond' by the protagonist in Milton's (1647) epic poem *Paradise Lost* provide another instance of the explicit use of pilgrimage as a literary device. It is interesting for this chapter, because of the way a natural landscape feature is used as a metaphor for Pilgrim's despair and depression. The journey that pilgrims undertake removes them from the restrictions and constraints that normally regulate their social interactions, in a way which is liberating. It allows them to reflect on their usual behaviour and relationships in a more detached way, helping to resolve problems that may seem difficult to tackle in their normal environment. The idea of spiritual journeys are also often used in ways which help people to contend with concerns and fears about death.

Journeys of a more secular kind are also used as literary metaphors for discovery that is important for wellbeing. For example, Gesler (2000) interprets Thomas Mann's novel *The Magic Mountain* partly in terms of metaphors for transitions to new knowledge. Addressing a more popular audience, Will Self's (2007) *PsychoGeography* also expresses at a more personal level the attraction of travelling on foot through places which bring back childhood and family memories.

Involuntary displacement is often more damaging to mental health. Refugees accounts of deportation or flight often emphasise the psychological impacts of loss, for example. Also, authors such as Young and Willmott (1957) and Fullilove (1996, 2005) have described how urban regeneration schemes, which aim to improve living conditions for residents in poor quality housing, may cause psychological damage. This may partly result from the break up of social ties in traditional neighbourhoods. Fullilove's critical analysis of the impact of regeneration also draws attention to the destruction of well-known buildings and streetscapes which had provided the settings for social interaction within communities, as discussed in Chapter 5. These were often built to a more idiosyncratic design, and on a more intimate scale than the more massive and impersonal architecture that has often replaced them in the form of high rise housing blocks and modern roads and shopping centres. Fullilove (2005) uses the term 'root shock' to express the psychological impact of this 'uprooting' of communities in American cities. Her arguments about the psychological and emotional bond to place echo ideas of Topophilia discussed in Chapter 2, and *Solastalgia* considered earlier in this chapter. She suggests that one's sense of self is partly defined in terms of the places where one lives and that people have a psychological bond to places, through emotional attachment, familiarity and self-identification which is disturbed by displacement, leading to distress arising from nostalgia, disorientation and alienation (Fullilove, 1996).

We see from these examples the symbolic as well as more tangible associations through which dynamic processes of changing places and movement through space and time relate to one's psychological state. The associations are reciprocal; one's mental state determines the view placed on imagined places, and the symbolic values of places may act upon one's mental state. We also see again in these examples how experience over the life course is incorporated into a person's view of transformation of place, displacement, origins and destinations.

Stigma, fear and phobias

To close this chapter attention is given to examples of research which puts a more sinister interpretation on the ways that symbolic aspects of space relate to one's psychological state. These include processes of social construction of stigma and fear as well as phobic responses and somatisation of distress, as they are expressed through imagined space and place. These issues are addressed from a geographical perspective in research on emotional geographies (Davidson, and Milligan, 2004; Davidson, Bondi and Smith; 2005)

Several of the sources already referenced in this book have shown how places, and the attributes of places, are often used in discourses which seek to stigmatise the people living there (e.g. Popay, Thomas et al., 2003). Further examples include a study by Takahashi and Gober (1997) showing how amongst a national survey sample from the USA, the stigma associated with mental illness seemed to translate into an unwillingness to tolerate psychiatric facilities in close proximity to one's home. McGregor (2008) analyses of 'abject spaces' through which Zimbabwean refugees navigate legal and social institutions that put in question their right to remain in the country and place them at a disadvantage in British society. The accounts of the interviewees reflect their feelings of anger and fear at their treatment, and are similar to the findings of other studies of the psychological health of refugees (e.g. Bhui et al., 2006; Warfa et al., 2006).

The image of something out of place may also be used to stigmatise certain social groups. Cresswell (1997) comments on the discriminatory language employed in government representations of the 'weed and seed' programme in the USA. Cresswell (1997: 335) cites a US congress document describing the aims of the intervention as being: "to 'weed out' crime from targeted neighbourhoods and then to 'seed' the targeted sites with a wide range of crime and drug prevention programs." Cresswell comments (pp. 335–6): "Using the 'weed' metaphor in the context of urban areas such as Los Angeles is not merely descriptive...[it]...referred also to the government's prescriptive goal of ridding problem areas of undesirable inhabitants (weeds)...This connotation of out-of-place people is attended by a host of other less obvious implications based on the characteristics of weeds. These out-of-place people may be viewed as weak but cunning, as reproducing quickly, as 'fugitives' always on the move... 'aliens' invading the proper order of the American city."

He goes further to suggest that this plays to ideas from social ecology of 'natural areas' of a city and 'cultures of poverty' which treat poor areas of cities as 'natural phenomena', rather than the outcome of socio-economic processes and suggest that certain social groups are 'culturally predisposed' to behaviours which explain their disadvantaged social position. This example therefore illustrates the use of imagery from the natural world to create a stigmatising image of particular local communities potentially damaging to their mental health.

Geographical research has also shown how fear of violence, crime, bullying and other forms of incivility may be experienced variably in different settings and by different groups such as women and men, younger and older people (Little and Panelli, 2005; Pain, et al., 2005, 2006 and Pain, 1997, 2006; Brownlow, 2005; Milligan, Gatrell and Bingley, 2005). These accounts show how individuals modify their use of space according to their perceptions of the risks of violence and crime (which may not correspond directly to the actual statistical risk). Geographies of fear depend partly on the reputation of certain places. Chapter 5 discussed the way that certain features of physical space are attributed special significance in this respect, including 'broken windows' and graffiti. Some accounts by residents of areas with a dangerous reputation find these features intimidating and worrying and outsiders may have fears about an area which are founded on its reputation rather than on direct experience.

In contrast, some people living in areas which have a reputation for social disorder consider that, being familiar with their surroundings they have no reason to feel particularly at risk (Popay, Bennett et al., 2003). Also, some of the supposed signs of incivility can also be read differently by certain observers and in certain settings. For example, for the graffiti artist, the image is interpreted as an expression of resistance to authority. Also, some graffiti have a less threatening aspect (see Figure 6.10) and do not convey any sense of threat. Thus the signs and symbols by which environments may be 'read' as 'safe' or 'insecure' are contingent on both the setting and on the characteristics of the individual observer. These examples all show how psychological wellbeing and the experience of mental illness are

Figure 6.10   Grafitti of an unthreatening kind

Note: Graffiti are often treated as indicative of environments that symbolise threatening environments and social disorder presenting a risk for mental health (see Chapter 5). However the impression may vary according to the content and location. This graffiti message, in pink, cursive script on a wall in a relatively secure area in Montreal, Quebec, conveys a 'warmer', more playful emotion and illustrates how graffiti may communicate and symbolise messages in ways that support psychological wellbeing.

closely bound up with social relationships involving power and control in places that can be characterised as 'spaces of dominance' or 'spaces of resistance'.

More extreme examples of individually specific psychological responses to imagined attributes of environment include experiences of obsessive compulsive disorders and phobias. A growing field of research in emotional geographies has investigated this aspect of imagined space and mental health (e.g. Davidson, 2003). Examples of studies on this topic include Davidson's (2000) interpretation of 'Agoraphobia' as a phenomenology of fear. She comments (Davidson, 2000: 653–4) that "Merleau-Ponty's existential analysis reveals that home is often the centre of our lived space. It constitutes the heart of many of our life-worlds and provides a relatively stable base for us to orient ourselves...Agoraphobics...are especially in need of this stability...[their] anxieties are generally associated with attempts to project themselves beyond the home, which is ordinarily the safe centre from which their lived space radiates". For 'Linda', the agoraphobia sufferer in Davidson's study, leaving home can be difficult. The nearby town centre is as far from home as she feels safe and even this may be difficult on a day when it is busy and crowded. Davidson suggests that Merleau-Ponty provides a framework and a vocabulary which help us to understand the agoraphobic experience. A historical analysis of 19th century interpretations of agoraphobia by Callard (2006) seems to suggest that experts at that period found it difficult to understand agoraphobia partly because they lacked these kinds of conceptual tools.

Another illustration is provided by Segrott and Doel (2004) in their account of Obsessive Compulsive Disorder (OCD). This condition causes the sufferer to be troubled by an obsession (an 'unwanted, intrusive, recurrent and persistent thought image or impulse') giving rise to compulsive behaviour that the person feels driven to perform (De Silva and Rachman, 1998; Baer, 2001). The severity of OCD is variable, but at its most severe is completely debilitating. Segrott and Doel (2004: 601) describe the experiences of 'Jane' whose obsessive thoughts centre on the idea that she will contract a fatal illness through contact with germs: "However contamination comes about, an object will remain contaminated until properly disinfected, a process which is often as much symbolic (and ritualised) as material". For 'Jane' the decontamination process involves "washing, cleaning, ordering, maintaining barriers between herself and sources of contamination". For Segrott and Doel (2004: 609), "A geographical analysis is fruitful because of the way in which obsessions and compulsions are embedded within the spaces and practices of everyday life and have the potential to transform them radically." They also (p. 597) situate this type of experience within wider debates about the "pathological nature of modernity and post-modernity as socio-spatial formations", which suggests a connection with discussion about risk and uncertainty in contemporary societies.

As suggested at the start of this chapter, geographical understanding of the complex interaction between individuals and their environment are challenged and advanced by research on emotional geographies and phobic or obsessive-compulsive states of mind. Also, by adopting a view which puts in question the Cartesian distinction between body and mind we can come closer to understanding the experience of these psychological conditions. Williams (2000) suggests that there is potential to consider health from a perspective on emotions and identifies key organising themes in terms of: reason vs. emotion; biology vs. society; the micro and the macro divide; and medicalisation vs. de-medicalisation.

Conclusions

The selective examples in this chapter have examined from a geographical perspective how imagined spaces, and 'real' places that are invested with socially constructed symbolism, can be associated with psychological health and wellbeing. The way we perceive, remember and interpret certain settings can be important for our emotional response to place. We have discussed here aspects of memory, including 'working memory' as studied by neuro-science, processes of recall and reconstruction of past experiences emphasised in psychological accounts by influential writers such as Freud, as well as culturally constructed 'folk' memories, traditions and myths which give certain places emotional significance. The importance of memory and imagined spaces underlines the need to consider mental health geographies in a temporal as well as a spatial framework. Freud's interpretation emphasises the way that both recent and more distant memories are revisited in dreams. The following comment by Campbell, underlines the way that memory works simultaneously with information gathered in the past as well as the present.

> Memory seems to be a way of spanning temporal perspectives, a description of the world formulated at one time used when describing it in another. This procedure assumes that there is a single temporal reality onto which all one's various temporal perspectives face, so that the judgements I make at one time can be the basis of knowledge at a later time. This characteristic of memory brings our ordinary reliance on it into conflict with any view which opposes realism about the past. (Campbell, 1994: 224)

From the perspective of complexity theory, this seems consistent with the idea of path dependence, through which one's sense of self and one's response to current environmental conditions are partly conditioned by memories of experiences built up throughout the lifecourse. The function of memory therefore resonates with some arguments about complex, relational geographies.

We have also noted that there is cultural and social variation in the way that we understand our relationship to natural and built environments. Many cultures that are less individualistic in outlook than the dominant social groups in European and North American societies have a world view which emphasises quite strongly the importance of natural landscapes and ecological relationships for psychological as well as physical wellbeing. Certain built environments also have particularly strong cultural significance for certain social groups. These perspectives underline the links between the relationships considered in this chapter and those reviewed in previous chapters. The symbolic and imagined attributes of landscapes are key to understanding psychological responses to different material and social settings.

Key learning objectives

This chapter aims to provide an introduction to research in geography and other disciplines that demonstrates:

- Variations in human perceptions of environment and the ways that these are interpreted socially;
- Connections between geographies of religion and spirituality and mental health geographies;
- Emotional responses to landscapes.

Introductory reading

For an striking account of savant perceptual abilities:
Tammet, D (2006) *Born on a Blue Day*. London: Hodder and Stoughton.

For a discussion of a culturally distinct perception of the links between dreams, landscape and spirituality:
Chatwin, B. (1988). *The Songlines*. London: Penguin.

For overviews of research on emotional geographies:
Davidson, J., and Milligan, C. (2004). Embodying emotion, sensing space: introducing emotional geographies. *Social and Cultural Geography*, 5(4): 523–32.
Davidson, J., Bondi, L. and Smith, M. (2005). *Emotional Geographies*. Aldershot: Ashgate.

For an introduction to geographical research on religion and spirituality:
Proctor, J. (2006a). Introduction: Theorizing and studying religion. *Annals of the Association of American Geographers*, 96(1): 165–8.

Chapter 7

Post-asylum Geographies of Mental Health Care: Spaces for Therapy and Treatment

Summary

This chapter draws on research in geography and related disciplines concerning 'geographies of care' to treat mental illness and promote mental health and wellbeing. A significant body of geographical work has focused on the social geographies of historical settings for mental health care, and particularly on the development of the asylum as the main setting for care of serious mental illness during the 18th and 19th Centuries. More recent deinstitutionalisation of psychiatric care has given rise to a new wave of geographical research on post-asylum care settings. This chapter considers geographical understandings and interpretations of various places where psychiatric medicine and other therapies are provided to people suffering from mental illness or psychological distress. Selected examples are used to illustrate how geographical perspectives have been used to interpret the blurred and complex interrelationships between these different types of care setting, as well as the material, social and symbolic attributes of specific 'care spaces' which contribute to their therapeutic properties. This chapter concludes with a discussion of how variations in patterns of provision and use of psychiatric services may be interpreted in terms of arguments about territorial justice in the distribution of services relative to population needs.

Introduction: geographical perspectives on care and treatment

Earlier chapters were concerned with geographical factors associated with states of mind and mental health. Here the discussion shifts to focus more specifically on geographies of *care and treatment*, concentrating on services that aim to care for mental illness and provide support and therapy to help people who are psychologically distressed. (In the final chapter the emphasis will be more particularly on maintenance and promotion of good mental health and wellbeing).

Research on geographies of psychiatric services have a long pedigree and some of the earliest research on geographies of health care, dating from the 19th century, concentrates on geographical variation in use of psychiatric facilities. An analysis by Jarvis (1866) examined the distribution of distances travelled to use psychiatric asyla in the 19th century. He classified geographical areas according to distance from each asylum, observed the numbers from each area coming to the asylum and calculated the rate of use, relative to population size for groups of areas in different 'distance categories'. He concluded that there was what we now refer to as a 'distance decay effect' in that more people came to the asylum from close by than from further away. This was an innovative approach and the 'distance decay effect' has sometimes been referred to as 'Jarvis' Law', since he found that it seemed to apply rather generally to several different institutions in the USA. Hunter and Shannon (Hunter et al., 1985, 1986) reconsidered his original information and argued that the 'distance decay effect' was variable for different asyla, suggesting that the distance patients were likely to travel would depend on factors such as ease of transport, admission policies of different institutions and differences in the cost of treatment.

As we will see below, questions about spatial proximity to psychiatric facilities and patterns of use continue to be of interest today. Research themes in the geography of mental health care have also developed and diversified over time to include a range of other topics. These include: varying views, in different societies and at different points in time, of what settings are appropriate for treatment of mental illness; the population characteristics in different localities that are associated with geographical differences in use of psychiatric care; concerns with the geographical pattern of provision of psychiatric care and how this relates to variation in estimated need for care in the population; how treatment for mental illness needs to be provided in different types of geographical area; whether settings for treatment are inclusive and sensitive to patients social and emotional needs as well as their clinical requirements; and how treatment settings relate to the quality of the care experience for service users. In the following discussion we consider examples of research in these fields, with a particular (but not exclusive) emphasis on research by geographers.

These topical debates in geographies of mental health care have been associated with some broader issues that preoccupy health geographers more generally. For example, Parr (2003), Wolch and Philo (2000) and Curtis (2004) trace developments in the discussion about geographies of mental health care through themes including: cultural, socio-political and economic perspectives on the social construction of illness and models of care for patients and the power relationships involved in treatment settings; concerns with issues of socio-geographical inequality, equity and social justice and ethics of service use; trends in collective consumption of mental health care provided by the state, and the ways that health care pathways and settings now reflect entrepreneurship and the commodification and commercialisation of health care. Geographies of health care have also widened in perspective, and rather than concentrating particularly on psychiatric inpatient facilities, there has been a shift towards research on care in a range of different 'spaces of care'.

As we will see in this discussion a number of different theoretical perspectives are also evident in this research. On one hand, we see research from a critical realist perspective on policies for health service provision, while on the other hand, interpretations from cultural and emotional geographies consider the individual's experience of health care settings. Both approaches have their strengths; Parr (2004: 246–9) considers arguments for a more 'critical' perspective and she identifies a number of strengths of critical geography. These include: its potential to be internationally relevant for health care planning and policy and to be usefully applied in mental health care practice; the interdisciplinarity of approaches, which, for example, often involve collaboration between medicine, geography and health economics; it is often theoretically and methodologically sophisticated and allows analysis of issues of scale and socio-geographical diversity. Reviewers such as Parr (2003) and Thein (2005) also demonstrate how more intensive, cultural, ethnographic and phenomenological approaches are important in research on health care. These tend to focus on specific case studies, so the findings are usually not generalisable, except in a theoretical sense. However, in their own way, they also provide important insights and understanding of social spaces and processes that are important for geographies of mental health care. They focus attention on the importance of geographies of affect for our understanding of people's psychological reactions to care settings and their experience of treatment. They also often give us a clearer impression than is evident from more extensive research of the feelings and desires that motivate human behaviour in seeking mental health care and using treatment, and the quality of the therapeutic experience.

The historical development mental health care: classic asylum geographies

The historical development of ideas about the suitable places in which to provide mental health care has been reviewed by geographers including Milligan (2000a), Philo (2004) and there is an extensive geographical literature analysing specific 18th and 19th century asylum facilities (Philo, 1989, 1995, 1997; Philo and Parr, 2000; Wolch and Philo, 2000). Here the particular focus is on the tensions and contradictions in attitudes towards mental illness that are revealed through these historical analyses of spaces of care for mental illness.

Before the 18th century, people manifesting psychotic behaviour were deemed deviant and 'sub-human' and left to the care of their family or an assortment of other institutional spaces of shelter or incarceration, which were not well designed for them. Little specific provision was made for them, except in a few cases, such as the Bethlem Hospital established in the 13th century, on the outskirts of the City of London (Philo, 2004: Chapter 6).

By the late 18th and early 19th century criticism was mounting of the conditions in those institutions that existed for people with mental health problems, and of the lack of provision in many other areas. A view was emerging of mental disorders as suitable conditions for therapy, and the 'moral treatment' movement developed. In countries such as the UK, North America and France this led to the idea that it was important to separate people with mental illnesses from the stresses of everyday life, especially in urban settings and that they were best cared for in specialised asylum facilities, in locations at a distance from city centres and in relatively tranquil, rural surroundings. At the same time, this strategy served the purpose of segregating mentally ill people from the rest of society, and placing social controls on their behaviour.

Ideas of retreat and refuge are reflected in the ideas put forward publicly at this period by the moral treatment campaigners who developed ground-breaking designs for asylum buildings and care regimes. In the UK, these included actions of Battie, Tuke and Browne (reviewed in detail by Philo, 2004), in France 'alienist' reformers such as Pinel (Coldefy and Curtis, 2009), and in the USA the work of Brigham, Todd and Wyman (reviewed by Hunter et al., 1986; Luchins, 1989; Whiteley, 2004.)

William Battie's (1758) *Treatise on Madness* is cited by Philo (2004: 462–3) as follows:

> madness…requires the patient's being removed from all objects that act forcibly on the nerves and excite too lively a perception of things, more especially from such objects as are the known causes of his disorder…for the same reason…the air he breathes should be dry and free from noisome streams…his amusements not too engaging nor too long continued, but rendered more agreeable by a well-timed variety…his employment should be about such things as are rather indifferent, and which approach the nearest to an intermediate state (if such there be) between pleasure and anxiety.'

William Tuke (and members of his family) were merchants and Quakers in the city of York, England, who were concerned to design an asylum based on principles of 'moral treatment' and Quaker values (for a detailed discussion, see Edgington, 1997; Philo, 2004: Chapter 6). Edgington (1997: 95) cites William Tuke's description of a suitable setting for an asylum as being "…in an airy situation, and at a short distance from York as may be, so as to have the privilege of retirement". This was in contrast to the pre-existing York Lunatic Hospital, located closer to the city centre.

Tuke's arguments concerning the York Retreat were incorporated into the 1815 *Report of the Committee on Madhouses*, which sought to reform psychiatric institutions of the day. Philo's (2004) detailed study traces the development of asylum facilities in England through the 18th and 19th centuries and shows how, in response to these new ideas about where to locate psychiatric treatment, many of those established in the late 18th and early 19th Centuries were built, as Tuke suggested, on sites which at the time were beyond the built up areas of major towns.

The vision of the 'moral treatment' movement could be considered as an early articulation of what makes for a therapeutic landscape for mental health care, incorporating physical, social and symbolic environmental elements that contribute to healing of mental illnesses. As well as stipulating a rural or semi-rural setting with access to tranquil and restorative natural landscapes, reformers like Tuke had a clear vision of the kind of interior settings that would be therapeutic and this was reflected in other facilities as well as the York Retreat.

Edginton (1997: 91) cites a contemporary description of an asylum in Wakefield, England in the mid-19th century, which noted the "..lightness, cheerfulness and agreeable temperature found within, the sense of adequate space, and the appearance of comfort, added to an extensive view of the surrounding country...". The design of rooms, furniture and windows in the York retreat, described at about the same time, was also considered 'homelike' (Edginton, 1997: 97) and Edginton's account of archival material shows that inmates were encouraged to participate in therapeutic activities such as gardening, sewing sports, music, dancing and outings arranged for patients. At the same time, there is evidence of an aspect of moral treatment which sought to impose social regulation and restraint on people being cared for in the asylum. For example, the social organisation of the York retreat emulated the social stratification of the wider society of the day, with separation of patients from different social classes, for example.

Furthermore, the principle of control of people with mental illness was never far below the surface. Even in the York retreat, Edginton explains that there is mention of doors that could be locked to restrain patients, and which were designed to open outwards to prevent patients barricading themselves in.

At around the same time, at the end of the 18th century, alternative models of institutional spaces for psychiatric care were also being put forward with a more pronounced emphasis on the carceral aspects of institutional psychiatric care. Figure 7.1 illustrates Bentham's (1791) idea for the *Panopticon* or 'inspection house', an architectural design put forward as suitable for all types of institution where inmates were to be subject to surveillance and control, including hospitals and what Bentham referred to as 'madhouses'. The design features a central surveillance point from which the interior of all the surrounding cells for inmates could be observed. The occupants of the cells could not see each other and would not be able to tell whether they were under surveillance or not. It was argued that inmates would therefore adjust their behaviour to the rules of the inspection house whether or not they were actually under surveillance at the time.

The original publication concerning the Panopticon shows that Jeremy Bentham (1748–1832) argued for this structure as an improvement on the kinds of carceral settings, such as ordinary prisons, that had hitherto been used to house people with mental illness and which had given rise to such concern among those arguing for moral treatment. In *Letter XIX: Madhouses* Bentham (1791 pp. 96–8) describes "...that strange and unseemly mixture of calamity and guilt; lunatics raving and felons rioting in the same room". He goes on to argue that in an 'inspection house' facility "...every vacant cell would afford these afflicted beings an apartment exempt from disturbance and adapted to their wants".

These 18th century visions of suitable settings for psychiatric care gave rise to a series of 20th century writings by Foucault (1967, 1973, 1977) who explored ways that social control is exercised in modern societies. His analysis of 'madness and civilisation' argued that as society developed a more 'civilised' understanding of psychiatric disorders it also refined ways to exert psychological as well as physical control over the behaviour of people with mental illness.

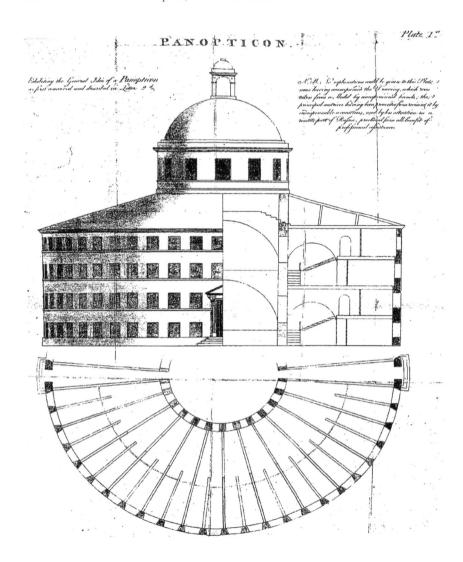

Figure 7.1   Jeremy Bentham (1791), Panopticon: or, the inspection-house. Containing the idea of a new principle of construction applicable to any sort of establishment, in which persons...are to be kept...and in particular to penitentiary-houses, prisons

Source: Eighteenth Century Collections Online, Gale Group. http://galenet.galegroup.com/servlet/ECCO. Reproduced with permission.

Foucault explored how the principles described in Bentham's *Panopticon* (and some aspects of the architectural design of the 'inspection house') were applied in psychiatric institutional settings. His other writings showed how this process was associated with the rise of medical power in clinical settings (Foucault, 1973) and he makes a similar analysis of regimes in prisons which sought to 'transform' individuals using the technical apparatus of disciplinary regimes (Foucault, 1977). Philo has extensively examined Foucault's work from a geographical perspective, emphasising in particular the ways that Foucault interprets spaces of power in institutional settings as reflections of the power relations in society more generally (e.g. see Philo, 2000 for a succinct statement).

Thus we see that from the 18th century onward, in most western countries, the idea of the asylum as a 'carceral space' is present in discourses about the organisation of psychiatric care, and that this is complemented repeatedly in the literature by the notion of the asylum as a 'refuge' or 'fortress', protecting people with mental illnesses against the risks of abuse, stigma or corruption to which they are vulnerable in the wider community and which might have an "unprofitable effect upon the patients' minds...even corrupting them with vicious habits..." (Foucault, 1967: 243, citing Tuke). These representations of psychiatric hospitals emphasise their character as 'spaces of power' as well as therapeutic paces and emphasise the interaction between the *material environment* of hospital buildings and facilities and the *social relations* within the hospital. We can also view psychiatric asyla as *symbolic* spaces, signifying the tensions in social constructions of people with mental illnesses, as sick and vulnerable or as 'deviant'; on one hand seeking to offer them a therapeutic and protective environment, but on the other hand motivated by a drive to subjugate, control and segregate them.

Post-asylum geographies of mental health care

Starting in the 1950s and 1960s, but gathering momentum in countries such as Britain and North America through the 1970s and 1980s, there has been a trend for psychiatric services to swing away from asylum provision, towards a deinstitutionalised model of care. This new model favoured treatment and support in community settings as far as possible, with the hospital continuing to serve a role as a source of care during periods of crisis in mental illness, but no longer aiming to make long term provision for psychiatric patients.

Geographers have taken a growing interest in these developments. For example, themed issues in the *International Journal of Health & Place*, introduced by Park and Radford (1997), Wolch and Philo (2000), and Philo (2005) have focused on 'post-asylum geographies, exploring a wide range of care spaces in community settings, such as out-patient clinics, consultation rooms for psychotherapy and counselling services, and more informal, social care networks including drop-in centres, and support groups. For a person with mental illness living in the community, their own home (if they have a fixed place to live) may also be a space for mental health care, for example, as treatment and therapy are delivered to clients at home. Here we consider some of the themes that emerge from these analyses as well as perspectives from other disciplines, demonstrating how material aspects of space and place interact with social and symbolic constructions in post-asylum geographies of psychiatric care.

Carpenter (2000) comments on development of mental health services under welfare capitalism since 1945, especially in the UK and the USA. Although he is not writing from an explicitly geographical perspective, his interpretation is relevant here for the way that it contextualises deinstitutionalisation within national health systems, presenting the trend as contingent on aspects of health policy prevailing in particular parts of the world. The process of closure of the asyla was associated partly with the development of new drug treatments which meant that it was no longer clinically necessary to keep people with mental illness confined to institutions. However, the processes causing the move to deinstitutionalisation were not limited to clinical arguments about the most therapeutic setting for treating mental illness, but were also influenced by social, political and economic factors.

Carpenter (2000) notes that not all countries in Europe followed the early trend towards deinstitutionalisation. He suggests that the change in psychiatric care was associated with policies which were particularly marked in the USA and UK at the time, emphasising retrenchment and economic reform of state welfare expenditure, driven by calculations of relative cost of asylum care vs. care in community settings, and linked to visions of health and welfare services in terms of market models emphasising diversity, choice and competition (e.g. reviewed by Mohan, 2002). Thus we may interpret the trend toward deinstitutionalised psychiatric provision partly in terms of nationally specific processes in the political economy and discussions about cost-effectiveness in the UK, North America and in Australasian countries. At the micro-level of particular institutions, there is also variability in the developmental paths of psychiatric care. For example Moon, Joseph and Kearns (2005) have examined the history of two specific psychiatric hospitals, one in Canada and the other in New Zealand. They show the different trajectories of these two institutions, over time, involving transitions between the state and the independent sector, and exemplifying how the 'institutional survival' of psychiatric facilities is intricately bound up with national and local policies and welfare economies, as well as the individual institutional history of each hospital.

Carpenter (2000) also emphasises that the process of deinstitutionalisation of psychiatric care in the UK and the USA was related to socio-political arguments about social control of people with mental illness and management of perceived risks associated with their behaviour. Some commentators considered that a motivation to shift care from asylum settings was the desire to reduce the degree of social control exerted over people with mental illness and integrate them in the wider community, rather than isolating them in institutional settings. However, an alternative interpretation discussed by authors such as Carpenter is that deinstitutionalisation represents an *extension* of social control over people with mental illness, beyond the asylum walls and into the wider community. We therefore see policies for community based care as closely linked to public perceptions of risk relating to mental illness and pressures for risk management in psychiatric care. These arguments link to those of Moon (2000) and Dallaire and colleagues (2000) about the representation of people with mental illness in terms of risk and 'dangerousness'. Dallaire et al. (2000) consider the social and clinical interpretation of 'danger to themselves or others' posed by people with mental illness that are bound up in legal decisions on the care of individual patients. There is a tendency for representations of mental illness to emphasise violence (either harmful behaviour directed to others or self-harm) as an attribute of certain mental disorders. This seems to be motivated by the desire to minimise risk as far as possible, in a situation where assessment of the risk of 'dangerous' behaviour is often very imprecise. This unfortunately results in public misconceptions about mental illnesses as generally 'dangerous', although in the vast majority of cases violent behaviour is not a typical feature of mental illness.

Furthermore, the professional interpretation of risks associated with mental illness seem to be quite variable and contingent on non-clinical attributes of people and their environment. Dallaire et al. (2000) concentrate on examples of legal hearings concerning the involuntary commitment to institutional care for people with psychiatric illnesses. They argue that the legal decisions seemed to rest on a constellation of factors including the person's social roles and context and not solely clinical considerations and estimates of the risk of harm. Moon (2000) also shows that discourses about the need for confinement are strongly influenced by individual, highly publicised instances of suicide or harm to others perpetrated by a few psychiatric patients who had been released from psychiatric hospitals. More generally, Moon concurs with Dallaire and colleagues concerning the way that the debate emphasises the perception of psychiatric patients as 'dangerous' and liable to cause harm. The public perception of a need to manage this risk results in a tension in health policy between arguments for deinstitutionalisation, countered by rhetoric concerning the need for confinement of certain individuals viewed to be a danger to themselves or others.

These tensions between requirements for risk management and confinement as against the desire to respect the rights and liberties of patients and integrate them in wider society, also help to explain the continuing attention to institutional settings in 'post-asylum' mental health care geographies. The discussion below considers how geographical research continues to focus on the changing role of institutions, while also extending to a much wider range of mental health care landscapes. These studies also emphasise the extent to which psychiatric care is used to extend social control over people with mental illness, in community as well as inpatient settings. This means that post-asylum geographies of psychiatric services can still be understood to a large extent as being geographies of *power* as much as they are geographies of *care*. This is a recurring theme in the examples shown below, drawing on evidence from community as well as hospital settings.

Another theme in post-asylum geographies concerns the fate of the sites of abandoned asylum facilities. After these hospitals have closed, research shows that they often continue to have symbolic significance for local communities, as well as for the patients who had sometimes spent quite long periods of time living there. Curtis, Gesler et al. (2009) discuss comments from a consultant in a psychiatric inpatient facility in London suggesting that for some chronically ill patients these hospitals had significance as a relatively stable feature in their otherwise insecure and unpredictable geographical experience, so that they tended to form a sense of attachment to the hospital building. The same idea is reflected in a study by Parr, Philo and Burns (2003), who researched the aftermaths of closure of a long stay psychiatric facility in Scotland and the perceptions of patients who had been treated there. Some of their informants talked about the hospital as a place that had been their home, and the accounts recorded that some ex-patients were still drawn back to visit the park around the old hospital. Other research has focused on the new uses made of the sites of closed asylum buildings. Examples include studies by Joseph, Kearns and Moon (2009) on the Kingseat and Seaview hospitals in New Zealand and the debates over how to use the hospital buildings after closure, showing that these were not only influenced by economic considerations, but also by the socially constructed significance of the sites, linked to memories held by local communities of the old asylum facilities. These studies highlight the symbolic significance of the asylum in social representations of psychiatric disorder and treatment.

New hospital spaces for psychiatric care: the 'complex re-birth of the clinic' as a therapeutic landscape

Although the deinstitutionalised model of care has restructured the landscape of psychiatric care, there are some similarities between the processes that we can observe in psychiatric care settings today, as compared with the classic studies of asyla discussed above. Foucault (1977: 246–7) argued that social discipline and control can be extended beyond the walls of an institutional *Panopticon* building, into wider society, by varying, mitigating or interrupting sanctions on inmates and allowing those who have reformed (or, in the case of patients, 'recovered') to be released into the community 'on license'. There seem to be parallels between this idea of the 'open prison' regime put forward by Foucault, and the deinstitutionalised 'matrix' model of psychiatric care described by Tansella and Thornicroft (1998) that is now widely established in many countries. As discussed above, this 'matrix' comprises services including hospital care and a range of community based services. Furthermore, some of the ideas about how to create therapeutic settings that were put forward in the 18th century still seem quite pertinent to present-day psychiatric care developments.

We have seen that one aspect of the community-based care 'matrix' is a revised role for inpatient services. However, for some people with mental illnesses, long term care institutions may continue to be their main place of residence. Wiersma (2008) provides a rare example of a geographical study of the geographical perceptions of people with dementia living in long term care homes. Also, in forensic psychiatry patients may be committed to hospital care for long periods. Nevertheless, most inpatient services are now provided in 'acute' (relatively short stay) psychiatric hospital wards where the care regime is designed to treat 'crises' in mental illness that make it impossible for patients to remain living in the community, and then to prepare them for a return to life outside the hospital. In various countries, including the UK, the emphasis on 'acute' psychiatric units has been accompanied by a restructuring or replacement of the buildings used to provide inpatient care for people with mental illness. For example, in the UK, psychiatric hospitals which often dated from the late 19th or early 20th century and were no longer fit for purpose have been upgraded or completely replaced under a government programme of investment in new public buildings, the *Private Finance Initiative* (PFI). The PFI involves collaboration between the commercial companies, who build and service the new facilities, and the government, which has committed to long term leases to occupy the buildings in order to deliver medical care. The PFI scheme illustrates how the national political-economic context of health care policy frames the development of psychiatric care in particular countries. This wave of new hospital building has also provoked a lively public debate concerning design. It is argued that previous generations of 'modern' hospital design were more concerned with clinical functionality than the more general wellbeing of patients, staff and other groups of people using the buildings. There have been calls for a closer consideration of hospitals as civic architecture, with social and symbolic significance as well as clinical functions, and some of the ideas discussed earlier in this book concerning material, social and symbolic aspects of space figure importantly in these debates. Reviews of these discussions, which relate to hospitals in general, as well as to psychiatric services, and are not restricted to the UK include: Moos (1997); Gross et al. (1998); Francis et al. (1999, 2001); Lawson et al. (2003); Waller and Finn (2004); Gesler, Bell, et al. (2004).

An example of the application of geographical ideas about therapeutic landscapes to the evaluation of the design of a new acute psychiatric hospital setting is reported by Gesler, Curtis and colleagues (Curtis, Gesler, et al. 2007; Gesler and Curtis, 2007; Curtis, Gesler et al., 2009) and is described in more detail in Box 7.1. This study took the form of a qualitative, post-occupation assessment of perceptions of a newly built psychiatric inpatient unit in London and involved discussion groups with a small group of users and staff. It emerged from this study that, without very specific prompting, informants discussed the physical, social and symbolic aspects of the new hospital in ways that evoked some aspects of the therapeutic landscapes framework. Their comments also relate to the themes discussed above concerning perceived attributes of a comfortable, protective and homely environment, as well as issues of risk management and control of patients' behaviours (Curtis et al., 2007).

Figure 7.2   Interior of a newly built psychiatric inpatient unit

Note: The photograph illustrates efforts to create a 'homely' and comfortable communal space in a newly built psychiatric inpatient unit in London, UK. The design is modern, with plenty of natural light, and features such as soft furnishings, pictures and potted plants.

Not all aspects of the design as planned had been realised in the final building and some issues in the design and use of the building had not been anticipated in the original discussions to agree on the hospital design. This highlighted one of the limitations of the PFI scheme, which required agreement on the hospital design at one point in time and was not always sufficiently flexible to allow variation of contractual agreements with the commercial developer over time as the NHS requirements for the hospital building evolved. It also emerged that the overall relationship to wellbeing of people using the new hospital building was importantly influenced by the complex interactions between different elements of the design (Gesler et al., 2007). It was only possible to evaluate the overall quality of the therapeutic environment taking into account these interrelationships. The authors argued that this approach to evaluation, while not a substitute for more systematic research, did complement studies which focus on links between specific aspects of design and clinical outcomes (e.g. Ulrich et al., 1984), or more structured 'satisfaction' surveys. This study also places more emphasis on non-clinical outcomes of hospital design than evaluation methods that have been put forward by National Health Service estates managers, which focus more on clinical

Figure 7.3   NHS staff viewing a full scale 'mock up' model of an individual patient's room designed for a new secure psychiatric facility being built in England

Note: This illustrates the safety and security features of modern psychiatric inpatient settings. Security features of the model room include: furniture of soft materials; doors, shelving, light fittings and handles which are recessed, smooth and sloping to avoid risk of ligature. The window (which in the real building would have a more attractive outlook) has a metal screen allowing the window to be opened, but making it impossible to climb through or to pass contraband items. More 'homely' features respecting patient's privacy and autonomy, would include coloured furniture, good ceiling height, natural light and fresh air, en suite bathroom, and provision for patients to display pictures and posters, to personalise the space.

functionality (Gesler, Bell et al., 2004). Some of the sorts of features that were long ago identified by Tuke and his colleagues in the moral treatment movement were still being emphasised in these new hospital settings, designed to encourage positive and supportive social interactions and a homely, attractive ambience (see Figure 7.2 and Box 7.1).

The rather holistic approach to considering design allowed informants to express comments which underlined the contextually contingent, complex and contested nature of hospital spaces. Some of the comments made stressed the importance of the interactions between different aspects of the hospital environment and the differences of view between diverse groups of participants concerning the value of some of the design features. For example, Gesler and Curtis (2007) report discussions concerning the problems of reconciling the need to keep windows securely closed, the decision to provide patients with south facing rooms so that they can enjoy plenty of natural light, and the lack of air conditioning in the new hospital (presumably to reduce costs and make the building more energy efficient). Individually, these might have been seen as advantages of the design, but given the high temperatures recently experience in the summer in London, these factors combined to produce an uncomfortable environment when the wards overheated. Nursing staff also discussed the impossibility of reconciling all the needs and preferences for hospital facilities for a diverse group of patients who varied in terms of diagnosis as well as age, gender and ethnic and faith group. Another example was the design of security features in the new hospital. There was also some disagreement between nursing staff and consultants over whether it was desirable to have nursing stations with a clear view of the corridors and vision panels in the doors to patient's rooms for nurses to be able to observe patients at risk of self harm, as well as features such as smooth door handles to prevent risk of ligature (See Figure 7.3). The arguments for such features were expressed in terms of the responsibilities of hospital staff for risk management, but there were also perceived drawbacks (which, interestingly were expressed especially by consultants), in that these features detract from a 'normal, homely' atmosphere, impinge on patient's privacy and liberties, and do little to encourage more relaxed social interactions between patients and staff. These discussions of the rather specific design features for security in psychiatric hospitals also demonstrated the special character of these care settings and emphasise their continuing role as spaces of surveillance and control of patients, reflecting some continuity with the characteristics of the classic *Panopticon* environment.

However, Curtis, Gesler et al. (2009) argue, in support of other research reported by Quirk et al. (2004, 2006), that the psychiatric hospital now plays the role of a 'permeable institution' or 'space of transition', which differs in some ways from the rather closed environments found in the old psychiatric asylum facilities. This permeability and connectivity with the community outside the hospital contributes to the complexity and unpredictability of processes in the psychiatric ward.

This emphasis on transition was reflected in various aspects of the design of this acute psychiatric inpatient unit and the care provided. The comments reproduced in Box 7.1, concerning the advantages of maintaining skills for independent living and links with their families and community outside the hospital show how the care model aimed to prepare patients to return to the community. The point was also made that although the aim was to make the hospital homely and comfortable, this should not be over-emphasised, as there was a risk that patients might become institutionalised if they became too settled there.

The location of this new hospital was in an urban rather than a semi-rural or isolated rural setting, and consideration had been given to facilitating accessibility to the surrounding community (See Box 7.1). Furthermore, patients were not

Box 7.1 **Perceptions of therapeutic landscapes in a psychiatric hospital: A post-occupation evaluation of views about the associations between design and wellbeing in a newly built psychiatric inpatient unit in London, UK**

This study explored the relevance of the 'therapeutic landscape' perspective as a framework to understand how the design of new psychiatric facilities related to the wellbeing of patients, staff and other users of a newly built psychiatric acute inpatient facility in London, UK. Information was collected from discussion groups and individual interviews with small, self selecting groups of informants from amongst the staff (including consultants and nursing and managerial staff) and people who had used the hospital as patients. The patients took part in a discussion at a site away from the hospital in the premises of a voluntary organisation, and were well at the time when they took part in the study. The research was scrutinised by the local ethics review board and it was supported by the Nuffield Foundation through a small project grant, as an academic study, independent of the Health Service. In all, ten staff and consultants and seven service users took part in the study, which is a small proportion of the numbers using the hospital buildings (though all staff and a larger number of users had been invited to take part). The findings should therefore not be interpreted as representative of the views of all those likely to be using the hospital, and though a range of categories of stakeholders were involved, it was not possible to include, for example, family members or other informal carers of patients at the hospital.

The discussion groups and interviews were unstructured. Participants were asked:

What specific features of [the hospital] (in terms of physical layout, activities, etc.) do you think are good for the well-being of users and staff?

What specific features of [the hospital] do you think are not good for the well-being of patients and staff?

The researchers interpreted transcripts of the discussion from a therapeutic landscape perspective (e.g. Gesler, 2003), in terms of aspects of the material, social and symbolic features of the hospital environment that were important for wellbeing. Selected extracts from informants' comments are reproduced below and the headings indicate the broad themes that the researchers attributed to these (for a more detailed explanation of these interpretations of these quotes, see Curtis, Gesler et al., 2007).

Material, social and symbolic aspects of the building reflecting respect and empowerment for people with mental illness were seen as important by the informants:

"…well when I was there the nurse was looking after me alright. The nurse was kind, was very nice to me, they look after me, they [were] very good to me" (User cited by Curtis, Gesler et al., 2007: 598).
"[The building is] new, and it's tidy and it's neat, and for me that's important because it reflects…what you think of the people, and what you think of the service users that you're helping" (Consultant cited by Curtis, Gesler et al., 2007: 599).

Aspects of design offering privacy and refuge were valued, and enhanced by the way the staff respected patient privacy:

"…I had my own room, so I could do what I wanted" (User cited by Curtis, Gesler et al. 2007: 599.)
"…they made some effort to allow female services [i.e. segregated provision for women]…a separate sitting room if they want to be away from the male clients for a while…we've got four…separate sitting rooms on the ward, so that allows people to…have a break from the staff if they want without having to move them to another unit" (Consultant cited by Curtis, Gesler et al., 2007: 601).

Patients also appreciated interactive spaces and positive, inclusive interactions with staff:

"…if they turned round to the patient and said, 'Well, what did you do before you came here?' That helps slightly. 'Is there anything we can do to help you do that?'"(User cited by Curtis, Gesler et al. 2007, 602.)

Natural light and a 'homely' setting within reach of the community facilities outside the hospital were seen as beneficial; helping to maintain positive mood, making patients comfortable and maintaining their skills for independent living:

"Working in a bright open place has much more positive effect on both staff and clients who are using the building" (Staff member cited by Curtis et al. 2007, 603).

"A little bit more homely and less institutional-like and more like somebody's house" (Consultant cited by Curtis, Gesler et al., 2007, 603).
"You're not going to be put on a big dormitory as used to be before. They can try to maintain that independence as [if] they were in their own home… so they won't lose that skill that they already have" (Staff member cited by Curtis, Gesler et al. 2007 604)
"where the unit is situated is…[close to]…shops and things like that….people can keep up the family contacts and take part in normal non-psychiatric related activities" (Consultant cited by Curtis, Gesler et al., 2007, 605).

Views about the benefits of the hospital design were not unanimous in all cases. Some aspects of the design were also identified by the informants as detracting from the therapeutic aspects of the hospital setting, or were more contentious. These included: lack of spaces for informal socialising; lack of support staff to accompany patients on trips outside the hospital and problems of access to the hospital by public transport; lack of properly equipped prayer and faith based activities; limited choice of food. There was also some disagreement between staff and consultants over the extent to which patients should be subject to observation by staff for security reasons.

confined to locked wards, though visitors could not enter the wards without permission. These features were seen as positive aspects of the design, in keeping with the community orientated care model and the desire to respect patient's autonomy as much as possible.

These aspects of design also produced some challenges, however. Given the shortage and high cost of land in an urban setting, the site of the hospital was quite small, lacking secure space for outdoor activities for the patients within the hospital grounds. Also, service users commented that the site was uncomfortably located close to a major road. An inner city location, in a deprived area with relatively high levels of civil disorder also has implications for the regime within the hospital. Staff commented on their role in trying to maintain a 'protected' space within the hospital. They discussed, for example, the need to prevent unwelcome infiltration from outside by drug dealers, or people carrying dangerous weapons (Gesler and Curtis, 2007; Curtis, Gesler et al., 2009). In the old asylum settings, disassociation from undesirable social influences was achieved by geographical isolation in rural settings. In modern psychiatric hospitals, that are physically located within the fabric of the city, alternative measures are necessary and the internal space of the psychiatric ward is increasingly represented and designed as a 'space of risk management' from a social as well as a clinical perspective. One respondent commented:

Real life isn't always nice and so if you want to reflect real life in a mental health institution then you get the unpleasant things as well. ...we don't want [a] 'real life' situation, you want an 'ideal' situation and a 'safe' situation.. (Consultant quoted by Curtis, Gesler et al. 2009, 345.)

Figure 7.3, from a different hospital, reflects similar issues (see note to Figure 7.3).

Curtis, Gesler et al. (2009) argue that these issues raise questions concerning the extension of 'clinical space' into the community. There are evidently challenges for health care providers in ensuring adequate connections between different parts of the complex psychiatric care matrix, extending between inpatient facilities and a range of settings outside. Hospital staff also sometimes seemed rather ill at ease with their role in trying to reintroduce patients to life in the community. For example, a lack of available staff to accompany groups of patients off the hospital site limited the number of such excursions that were possible, and it was suggested by some informants that voluntary organisations in the area were better placed than health service staff to offer access to offsite activities.

Although, as discussed below, there is evidence that clinical and social control of psychiatric patients does extend into community settings, it may do so in a more partial way than was possible in the old long stay institutions. Psychiatric wards now function more as permeable spaces of transition. The difficulties of managing the permeable interface between the hospital ward and the surrounding community mean that the significance of the 'fortress' and 'refuge' function of acute psychiatric wards, providing a 'protected' environment for patients, now seems very prominent in some hospitals today, featuring as strongly as the 'carceral' control and restriction of patients, which attracted so much attention in 'classical' geographical research on the asylum. Curtis, Gesler et al. (2009) argue that some aspects of complexity theory may therefore be useful to conceptualise these developments, well as Foucault's theoretical model of the *Panopticon*. They refer to a 'complex "rebirth" of the clinic', to signal that elements from both theoretical frameworks may help us to understand the material, social and symbolic features of therapeutic spaces on psychiatric wards in Britain today.

After the asylum: Exclusion and 'drift' or integration and therapy?

A number of studies of post asylum geographies are concerned with therapeutic services provided in settings outside hospitals and the experiential geographies of people with mental illness who live independently in the community. Some community care settings are medical services such as general practices, community psychiatric nursing services, and ambulatory clinics. However, many others are non-clinical and they include a diverse assemblage of services and therapeutic spaces including counselling services, various occupational therapies and drop in centres, lunch clubs, social clubs, and so on. This makes the 'matrix' of mental health care in the community very complex.

As several geographical researchers (e.g. Conradson, 2003, Bondi, 2005) point out, these settings are often difficult to categorise in terms of the type of service that they offer, which often spans between clinical or non-clinical therapy, social care and self care. From an organisational perspective, these services are often located ambiguously between: formal public services, which may fund parts of their activities or refer clients to them; voluntary, charitable or faith-based organisations which may help to run the services and provide venues; and services that operate quasi-commercially if clients pay a fee or a contribution to costs of the therapy they receive.

The socio-geographical experiences of people with mental illnesses in post-asylum settings might be expected to differ from those found in asylum facilities, partly because of the greater complexity and diversity of the physical settings where community care is provided and also because some aspects of the social relationships and the social constructions of mental illness prevailing in non-medical, non-institutional settings differ from those observed in the asylum.

For people receiving psychiatric care for serious mental illnesses while living in the community, the clinical aim is to provide a care plan for the patient, offering appropriate medical and social support services. However, in practice, clients often do not have clearly defined pathways, planned by psychiatric referrals, through all of the range of 'care spaces' which they occupy in their daily lives. Also, some people experiencing psychological morbidity or distress will not be in contact with medical services. Thus individual's use of therapeutic settings in the community may be idiosyncratic, variable and relatively autonomous. Knowles (2000) and Pinfold (2000), for example, explain how mentally ill people, many of whom were homeless, reported time spent in a number of different 'safe havens' that they could use to obtain food and shelter, including services operating by different agencies, on different days of the week and in diverse places in the city. Some of these safe havens were specifically provided for vulnerable members of the community at high risk of mental illness, while others were ordinary commercial or public amenities such as cafes, parks, or libraries. Townley et al. (2009) have extended this approach, using mixed cartographic methods to record the action spaces experienced by people with mental illness living in two counties of South Carolina. They collected self-reported descriptions and cartographic representations from their informants as well as computerised cartographic measurements of the spaces in different parts of their community that the study participants made use of and considered important (see Box 7.2 and Figure 7.4).

Some accounts have been quite pessimistic about the degree to which a community care model for psychiatry results in social integration for mentally ill people into the wider community. They argue that there are continuities between 'asylum' and 'post-asylum' geographies, such that people with mental illnesses continue to be subjected to discipline, surveillance, social stigma, and segregation. For example, Dear and Wolch (1987), in their classic analysis of the impacts of deinstitutionalisation on the geographies of people with serious mental illness in North America, gave an account which stressed processes through which people

Box 7.2   The action spaces of people with mental illness in South Carolina: A study by Townley et al. (2009)

It is argued (see Chapter 4 of this book) that social and psychological integration into local communities is important for mental health of most people, and it is considered a key aspect of care and support for people with serious mental illness who are living in the community. This study by Townley and colleagues is an example of use of mixed geographical methods to study 'spaces of inclusion' that were important for integration for 40 people with serious mental illnesses who were currently using mental health services and were living independently in the community (not in sheltered housing or hostels), in two areas of South Carolina, USA. The aim was particularly to collect information on the various geographical settings that provided them with points of social and psychological integration into their local community and support for independent living, which we might consider as 'spaces of care' in the broadest sense. Townley et al. (2009: 523) explain that after introducing the participants to the study they requested them to 'draw the places that are important to you'. Participants drew sketch maps or other representations of these places and were asked about how they used these sites, why they found them important and how much of their time they spent at different locations. Information was collected to allow the sites of importance to the informants to be located geographically and mapped using a Geographical Information System (GIS). The researchers used the GIS to draw estimates of the areas defining each person's normal daily 'action space' where they typically spent their time. The study also collected survey measures of sense community, life satisfaction and recovery from mental illness, and measures of perception of belonging and acceptance into the community. The researchers noted that while some action spaces were relatively easy to plot on a map, being rather concentrated in a single area, some respondents had complicated action spaces involving concentration of activity in two or more rather separate areas. Thus the GIS definitions of action space were quite approximate. However, allowing for these limitations, the researchers showed that people with rather extensive action spaces, who spent time in places that were more widely distributed and further from home, tended to have better life satisfaction and show better levels of recovery. On the other hand, those who tended to spend most of their time in an area close to where they lived had a stronger sense of community. The study also showed that the participants often perceived their home as the place that was most important to them, and many of them also identified spaces for social activities and leisure, and for activities of daily living, such as shops and restaurants, as being important as well. Townley et al. (2009: 530) comment that by studying the context of the person's daily experience in these ways one can identify the physical dimensions of integration in the community and start to understand the settings that are important for social and psychological integration.

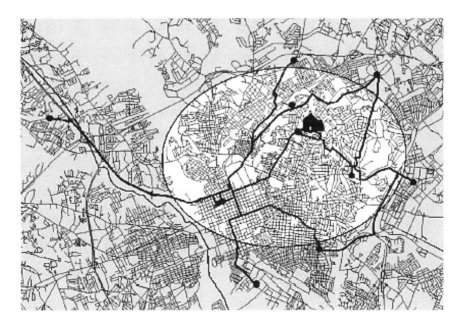

**Figure 7.4   A representation of the typical daily action space of a
participant in a study by Townley et al. (2009) of social
integration of people with serious mental illness living
independently in the community in South Carolina, USA**

Source: Townley, G., Kloos, B. and Wright, P. (2009). Reprinted from *The Lancet* with permission of Elsevier.

Note: The diagram is based on information from the participant about the places that they used regularly and that were important to them (paths marked by heavy lines on the map).

with mental illness tend to congregate in relatively deprived inner city areas with established services for people with social and psychiatric difficulties, which they termed the 'service dependent ghetto'. This process of selective 'sorting' of mentally ill people into certain geographical areas is sometimes expressed by the idea of 'drift'. As already discussed in Chapter 5 of this book, 'drift' may result from selective patterns of migration by people at greatest risk of mental illness towards certain types or area. It may also be associated with a relatively strong tendency for them to remain in poor inner city areas once they are there. Most studies of 'drift' effects focus on people who have already been diagnosed with serious mental illnesses. (See also discussion in Chapter 5.) The term 'drift' does not imply a socially neutral set of processes, and nor is it the outcome of the individual's free choices; it is the outcome of a combination of social and economic exclusion, stigma and disadvantage, as well as local concentrations of psychiatric care services in certain parts of the city. In addition to Dear and Wolch, other accounts by authors such as Smith (1978), Sibley (1995), Takahashi and Gaber

(1998), Wolch and Philo (2000), Parr (2008) have examined the significance of social processes of stigma and discrimination which lead to socio-geographical exclusion and segregation of people with serious mental illness into certain marginalised settings.

Other evidence further demonstrates how psychiatric service provision and use are closely bound up with the geographical distribution of people with mental illness in urban settings. Research by DeVerteuil et al. (2007) and Lix et al. (2007) discussed in Chapter 5 supports the idea that there is selective migration towards inner city areas on the part of people with schizophrenia and that individuals with the greatest levels of residential mobility make most use of psychiatric services. As mentioned in Chapter 3, Warfa et al. (2006) also discuss the high levels of residential mobility among Somali refugees in London, and comment on the stresses associated with housing insecurity as well as the problems of providing continuity of medical care for these very mobile individuals. These factors may contribute to the risks of psychiatric illness for these migrants.

At the population level, the tendency for people with mental illness to be relatively concentrated in poor areas of cities is reflected not only in geographical differences in mental health (as discussed in Chapter 5) but also in the pattern of use of psychiatric services. In an ecological study of patterns of psychiatric hospital admissions, Almog et al. (2004) demonstrated that zipcode areas of New York City with the greatest levels of socio-economic deprivation also had highest rates of hospital admissions for psychiatric conditions (See Figure 7.5) . The authors point out that this finding is not in itself unusual, since the pattern is replicated in many other studies from cities around the world (e.g. Schweitzer and Kierzenbaum, 1978; Dean and James, 1981; Thornicroft et al., 1993; Dekker et al., 1997; Maylath et al., 1999; Curtis, Copeland et al., 2006). The more original aspect of the research reported by Almog et al. (2004) is that the relative concentration of psychiatric admissions in poor areas had increased from 1990 to 2000. They explored various possible reasons for this trend, and reject the hypothesis that poverty might have intensified in the poorest areas over the period, concluding that the most likely reason was that changes in policies for funding and managing psychiatric inpatient care had produced clinical incentives for an increase in admissions. This finding underlines the complex and dynamic relationships between population characteristics and health service organisation that together affect the geography of mental health service use, and lends empirical support to theories postulated earlier by authors such as Dear and Wolch (1987).

It should also be noted that this pattern of progressive inner city concentration of people being treated by psychiatric services for mental illnesses does not provide a complete picture of the geographical experience of psychiatric service use. For example, some authors (Milligan, 1996; Jones, 2000) have pointed out that homes for small groups people with mental illnesses are not always concentrated in inner city areas in Britain. Furthermore, Philo and Parr (2004) consider migration trends among people in fragile mental health which oppose the trend towards concentration in inner cities. They report on a qualitative study in rural areas of

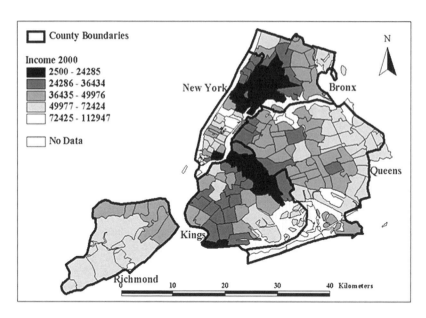

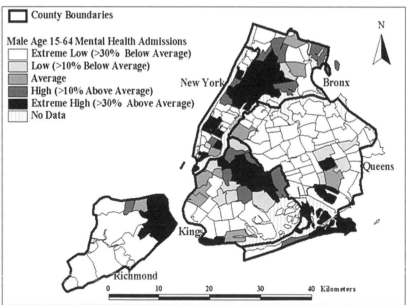

Figure 7.5   New York City in 2000. Zip code areas grouped in quintiles
              by income and Standardised Admission Ratios for psychiatric
              hospital admissions (Darkest areas have lowest income/ highest
              SAR)

Source: Almog et al. (2004). Reprinted from *The Lancet* with permission of Elsevier.

Scotland which records migration into the area by people who are seeking a more tranquil environment, away from urban stressors, and their findings show that some of these incomers were partly motivated by the perception that this would be more therapeutic for mental illnesses that affected them.

Other approaches in post-asylum geographies focus more on the micro level processes in specific 'care spaces' in community settings. These interesting here particularly for the way that they emphasise the social and symbolic aspects of the therapeutic environment they offer. Some aspects of these accounts suggest that these therapeutic settings control and restrict individuals. Parr (2000), shows for example, that in community centres for people with severe mental illness, service users themselves, as well as service managers, may take action to curb behaviours by other users that they consider unacceptable. In some respects, this echoes Foucauldian ideas about imposition of 'self-discipline' through processes of social control. On the other hand, some aspects of these accounts emphasise rather more positive socio-geographical experiences of mental illness and provide accounts of therapeutic settings that seem supportive and inclusive. Parr (2000) also describes how within community centres she observed, there was a degree of tolerance of behaviours which might seem odd or unacceptable in general public environments. In his analysis of a philanthropic 'drop-in' centre in a mid-sized British city, Conradson (2003), argues that these are often quite flexible spaces, which do not correspond to the kind of 'Panopticon' model proposed by Foucault to describe psychiatric asylum facilities. Although some of the users were having social and psychological difficulties, the facility was not intended to provide formal psychiatric care, but rather to provide a point of social contact and support for people living in the community. Conradson's account shows how the space was used variably by different individuals. He also explores the relevance of the therapeutic landscape model to explain these spaces. Different features of the drop-in centre seemed to be beneficial for different individuals. This finding reinforces other work by Conradson (2005), showing that the therapeutic effects of a particular setting may depend not on some universal experience, but more on the variable interactions between each individual person and the environment.

Parallel arguments are put forward by Bondi and colleagues, who have published a series of papers considering psychotherapeutic practice and consultation settings from a geographical perspective. Bondi (2003) discusses the nature of talking therapies, particularly counselling. She contrasts approaches based on Freudian analysis, where the therapist has a more controlling role, with those proposed by Carl Rogers in which the emphasis is more on empowering a person to identify and address themselves the issues for which they are seeking therapy or advice. Bondi and Frewell (2003) and Bondi (2005) make interesting links between these 'person-centred' approaches to therapy and ideas in emotional and feminist geography that privilege the position of the informant in research and focus on personal feelings and social spaces defined in terms of interpersonal relationships and interactions. These analyses emphasise the interactions between material and symbolic spaces involved in counselling therapy. Bondi (2005: 444) argues that the counselling process involves finding a way to represent or symbolise one's feelings in order to make them more accessible to interpretation and comprehensible. She states that it is "not accomplished by agents acting individually but is intrinsically relational" in the sense that it relies fundamentally on the relationship formed between the analyst and the patient and the interactions that take place between them.

Bondi and Frewell (2003: 528) undertook interviews with around 100 people providing counselling services in Scotland. Their study includes an exploration of how spatial language and imagery frequently figured in the accounts of their informants, as well as in the language of psychotherapy more generally. For example, frequent use of metaphors invoking spatial mobility and temporal dynamics were used to express the processes of 'transference' whereby memories of experiences throughout the lifecourse impinge on one's present mental state, and the aim of therapy to help a person to 'move on' from a difficult or distressing experience. Also, emphasis was placed on the idea of creating a 'safe' and 'confidential' 'space' in which to explore sensitive and private issues. This was reinforced by the choice of a 'material space' for consultation, often carefully selected to be separate from other places that the person receiving counselling occupies in other parts of their daily life, as well as being comfortable and relaxing. Bondi (2005) points out that the 'self disclosure' required in counselling involves 'crossing a boundary' of privacy to share personal information with a counsellor who is beyond one's immediate social circle (especially the family) and that this may be seen as a form of 'transgression' which is liberating for the client allowing freedom of expression of secrets or ideas that might be seen as socially unacceptable or difficult to share with people who are socially close. This 'boundary crossing' is also expressed spatially, for example in the setting chosen for counselling, which is normally separate from the home, place of work, or other parts of the normal social action space. The precise demarcation of the time and place of a counselling meeting marks it out as a distinct social and geo-temporal space. The walls of the consulting room are presented as symbolic of the confidential relationship of client and counsellor.

Other forms of therapy are centred on activity rather than talking and these have also been interpreted from a geographical perspective by Parr (2006), who builds on ideas from Probyn (1996), about 'belonging', Parr considers the importance of arts projects as social spaces that are not clinical, and which for some participants provided a place of social stability and belonging. She also discusses how participants viewed their presence in these spaces as transient and dynamic, being a step towards more general social integration. She also explores the idea that artwork may help to provide a reassuring sense of 'boundedness' (of a finished piece of artwork, for example) as well as providing alternative method of communicating other than ordinary conversation which may be too difficult for an individual who is very distressed.

Other forms of therapeutic space may be more explicitly orientated towards reintegration into wider social and economic structures of society. An analysis by Philo, Parr, and Burns (2005) examined Training and Guidance (TAG) units in the Scottish Highlands, aiming to help people with mental illnesses back into work. They found these units were represented as places of 'safety, attachment and belonging' and as 'in between' spaces where social and economic rationales and activities complemented each other. They also considered tensions created between these representations and the need for 'progression' towards mainstream employment.

Several aspects of these accounts of community based services parallel the points above concerning psychiatric hospitals as 'transitional' spaces. They share similar characteristics associated with the potentially contradictory requirements of these care spaces. On one hand they aim to provide a safe, protected space where people with mental illnesses can feel more confident that their illness will be sympathetically understood and they can undergo therapeutic activities and experiences to help them to recover good mental health or to manage better with chronic mental illness. They constitute an alternative social environment which, in contrast with wider society, does not stigmatise and marginalise people with mental illnesses and offers them stable spaces of attachment and structure to help them establish reassuring patterns of organisation in their daily lives. On the other hand, most of these care spaces are also designed to help people, as far as possible, to become better integrated and rehabilitated to independent living in the general community, so that there is a constant emphasis on *transition* out of these therapeutic spaces. This may be beneficial for some users, but for others the requirement for transition to independent community life may be unsettling and may seem to undermine the other benefits they gain from these therapeutic care spaces. It also challenges service providers who try to balance these requirements.

Informal Carers

The geographical research reported above is focused almost exclusively on services which are provided by organisations, either statutory services or voluntary sector agencies. Relatively little geographical research has focused on the role of informal carers, especially family members, looking after people with mental illnesses in their homes and neighbourhoods (though there are some notable exceptions such as Milligan's research on dementia care, discussed below). This is a gap in the geographical literature, considering the significant caring role that many family members undertake for relatives with mental illnesses, particularly serious long term illnesses such as dementia or schizophrenia. Georges et al. (2008) found in an international survey of carers in five European countries that among people looking after someone with late stage dementia the average time spent caring exceeded 10 hours a day. Goodwin and Hapnell (2007) describe informal carers' roles in looking after people with mental illness as 'pivotal' and suggest ways that community psychiatric nurses may be able to recognise this role more effectively and enhance the level of participation for informal carers in structuring the overall treatment process. The context of formal care for mental illness is variable from place to place, which will affect the burden placed on informal carers. The indications are that statutory services provide rather limited support to informal carers who take responsibility for people with chronic mental illnesses. (Indeed, it seems likely that statutory services are more particularly targeted to people with mental illness who live alone or are homeless and do not have a regular informal carer.) In the survey by Georges et al. (2008) of people providing informal care to Alzheimer suffers, less than a fifth of carers considered that the support from statutory services for people with dementia living at home was good and many were critical of the low levels of home care support services they received. Jeon et al. (2005) also found from a review of research that respite care for people with serious mental illness, which can temporarily relieve the pressure on informal carers is generally considered insufficient. Zechmeister and Osterle (2007) report that the health and welfare system in Austria is not well adapted to supporting long term informal care of people with mental illness. Roick et al. (2007) showed that the burden of giving care to people with mental illness seemed heavier for informal carers in Britain than in Germany and they suggest that this reflects differences in the organisation of support services.

Informal sector organisations may in some ways be better adapted to support carers than the professional care services. Milligan (2000b) discusses the ways that voluntary sector agencies support informal carers and the importance of voluntary agencies in supporting informal sector care. She also reports (Milligan, 2005) from a qualitative study in New Zealand of informal carers' roles in respect of a sample of older people who were living in residential care homes, many of whom were affected by dementia. The study showed that many informal carers continued to spend a significant amount of time providing a range of care, even within the care home setting. She also describes the concerns of carers about the transition of the main place of care provision, from family home to a care home, with all the attendant implications for their caring responsibilities and roles. At the other end of the age spectrum, although clinical models of treatment for childhood mental health problems often emphasise a 'whole family' approach to treatment, it seems that the caring role of parents is often ignored or underestimated by professional clinicians. Harden (2005) reported from a qualitative study of parents caring for children with mental illness about their feelings of frustration and helplessness over the difficulty of gaining adequate knowledge of mental illness to carry out their caring role and the failure of formal medical services to recognise and value their role. The wider social context determining the social construction of mental illness is also likely to be important for informal carers. The public response to behaviours associated with mental illness or disability can stigmatise carers as well as the person displaying the behaviour. Power (2008) in a study of carers with responsibility for young adults with learning disabilities discusses their experience of such signs of social stigma and social rejection.

Mental illness may also be caused by the caring role. Authors such as De Mattei et al. (2008), Campbell et al. (2008) and Van Den Wijngaart et al. (2007) have also examined the burden of care and shown how the stress and worry of the caring role and its impact on other activities in life can produce psychological distress in the carers themselves.

The informal personal caring role therefore raises a number of interesting issues from a geographical perspective, and although individual carers looking after people with mental illness have rarely been specifically the focus of research, there is a larger literature on informal caring more generally, as well as the site of informal care, which is often undertaken in a domestic setting. Dyck et al. (2005) and Yantzi and Rosenberg (2008) have examined the issue of care provision within the private domestic space of one's home. In some respects the home represents an ideal site for care provision, offering privacy, security and intimacy. However, there are also interesting issues concerning the 'invasion' of private domestic space by caring activities and the burden of care for informal family care givers. Lawson (2007) also comments on the ways that caring relationships affect the social relationships between the carer and the person cared for. Social structures such as health and social care systems are important but wider scale aspects of society also influence the context for informal care. For example, the strongly gendered role of informal care partly reflects social expectations for women's roles. Smith (2005) has also commented on how wider structural factors such as housing markets impinge on the potential and limits of possibilities for care in domestic settings. Furthermore, it would be misleading to assume that informal caring roles are limited to home spaces and solely provided by individual family members and friends. Lawson (2007) discusses a wider geography of caring roles, extending over longer geographical distances. Lingler at al (2008) in a study of caregivers for people with a range of illnesses found that this informal caregiving role is often quite complex and caregiving may be reciprocal. Thus elderly couples may provide care for each other, some caregiving is provided by a constellation of family members and other caregivers rather than a single carer, and some people who receive informal care, may themselves provide care to a third person.

Thus the research available suggests that the informal care giver's role in care and treatment of mental illness is in many cases crucial, and is bound up in quite complex ways with the contexts in which caring takes place. This caring role is contingent on a number of aspects of the local and more general socio-geographical context. At the wider scale, these include the roles of key social institutions such as health services, the organisation of informal sector provision, and general social attitudes towards mental illness and expectations about who should undertake informal care. For individual carers, a complex web of local processes will help to determine the nature and burden of their role, including the health and dependency level of the person they care for, the material and social environment of the domestic space where caring takes place, the relationships formed with statutory and third sector services, and personal relationships with the person cared for and other key individuals in their social circle.

Conclusion

This chapter has reviewed examples of a large field of research which has shown how in high income countries a post-asylum 'matrix' of settings for care and treatment of mental illness has developed, characterised by increasing diversity and complexity. Many of these accounts of post-asylum therapeutic services provide illustrations of the significance, for people who are mentally ill or distressed, of socially constructed and imagined spaces, as well as material spaces created in these care settings. It is possible to understand these post-asylum spaces of care in terms of the theoretical perspectives emphasised in this book. The research reviewed here provides further instances of therapeutic benefits for psychological health derived from the interaction of material, social and symbolic aspects of particular places, as anticipated by the therapeutic landscapes perspective. Several authors stress the continuing relevance of analyses of social relations in post-asylum care settings and the importance of understanding power relationships between the different groups of service users and providers that affect the therapeutic experience. We might also consider that people and other, inanimate actors, such as particular service locations, artifacts and therapeutic practices, interact in the kinds of complex socio-geographical spaces that are described by Actor Network Theory. Furthermore research on post-asylum geographies emphasises the dynamic and emergent aspects of mental health care systems, which evolve over time in response to changing ideas about the nature of mental illness and appropriate treatment strategies, but are also contingent on past service configurations in particular national and local contexts, as well as the existing knowledge and practices of the actors involved.

Much of the discussion here has focused on the ways that individuals experience various care and treatment settings. However, this chapter has also begun to demonstrate how at the level of whole populations and within national health systems, the provision of psychiatric and social care interacts dynamically with the geographical patterning of populations who need to use these services and the wider context of the communities where they live. The examples considered here suggest, for example, that access to therapeutic settings and 'safe havens' in the community is important for the wellbeing and social integration of people with mental illnesses. These findings are important for issues of access to services and equity in service provision which are given further attention in the next chapter.

Key learning objectives

This chapter aims to provide introductory knowledge of geographical research on:

- Development of models of psychiatric care and how this relates to geographies of psychiatric services;
- The interplay of material, social and symbolic aspects of environment in

settings where care is provided for psychiatric illness and psychological distress;
- The potential for geographical theories and techniques to be applied in planning mental health services.

Further introductory reading:

For a summary of some of the main lines of development of mental health geographies discussed here and further references:

Wolch, J. and Philo, C. (2000). From distributions of deviance to definitions of difference: past and future mental health geographies. *Health & Place*, 6(3): 137–57.

For a very comprehensive review of the development of psychiatric asyla in the 18th and 19th Centuries, especially in England:

Philo, C. (2004). *A Geographical History of Institutional Provision for the Insane from Medieval Times to the 1860s.* Lewiston, NY: Edwin Mellen Press.

For interesting discussions of geographical experiences of providers and users of psychiatric services:

Bondi, L. (2005). Making connections and thinking through emotions: between geography and psychotherapy. *Transactions of the Institute of British Geographers*, 30(4): 433–48.

Parr, H. (2000). Interpreting the 'hidden social geographics' of mental health: ethnographies of inclusion and exclusion in semi-institutional places. *Health & Place*, 6(3): 225–37.

Conradson, D. (2003b). Spaces of care in the city: the place of a community drop-in centre. *Social & Cultural Geography*, 4(4): 507–25.

Curtis, S., W. Gesler, et al. (2009). New spaces of inpatient care for people with mental illness: A complex 'rebirth' of the clinic? *Health & Place*, 15(1): 340–8.

Chapter 8
Place, Space and Population Mental Health

Summary

This chapter takes a public mental health perspective to address the question of how mental health might be improved at the population level. Particular consideration is given to the significance for public mental health of the geographies of mental health reviewed in this book.

The following discussion concentrates especially on strategies aiming to: prioritise good mental health and wellbeing of the whole population as an aim for sustainable societies; enhance social and territorial equity in the availability of services for treatment of psychiatric disorders to make them responsive to the needs of individuals in different geographical contexts; increase sensitivity to the relationships between population mental health and the material, social and symbolic complexity of environments, including closer attention to the mental health impacts of policies and interventions outside the medical sector.

The book concludes with some comments about the current and future agenda for geographical research into the complex processes relating to population mental health.

Introduction: the idea of population mental health

To conclude this overview of relationships between space, place and mental health this final chapter addresses the challenge of how to improve mental health for the population as a whole. A population mental health perspective encompasses, but extends beyond the provision of medical services to treat mental illness and distress. It places emphasis on strategies to promote good mental health in the population as a whole, as well as treatment and healing of mental illnesses and psychological distress for individuals. This chapter particularly considers the significance for population mental health of the geographies of mental health reviewed in this book.

The following discussion concentrates especially on the following approaches to promotion of mental health in the population as a whole:

- prioritising good mental health and wellbeing of the population as a central aim of health policy internationally;
- enhancing social and territorial equity, for geographically defined populations, in access to mental health care and treatment;

- sensitivity to the relationships between public mental health and the
  material, social and symbolic complexity of environments, which requires
  consideration of the mental health impacts of policies and interventions
  outside the medical sector.

In order to support this agenda into the future, new developments in geographical
research will be needed, to improve our understanding of the complex processes
involved. This suggests new directions for research in geographies of mental
health, discussed below.

Population mental health as a global priority

A number of reports and statements from the World Health Organization reflect
a concern at the global scale to emphasise the importance of mental health as a
crucial dimension of human health, and to take a view which considers mental
health at the population level, as well from a medical perspective concentrating on
individuals with specific psychiatric disorders.

In 2001, the World Health Organization published in its *World Health Report*
for that year an analysis of the global situation relating to mental health. This drew
attention to the significance of mental health as an aspect of human health and the
burden of mental illness in populations internationally (discussed in Chapter 1 of
this book). It made a number of recommendations for 'minimum actions required
for mental health care' in different parts of the world, which varied according to
the national resources available for health care generally and the current state of
mental health services in each country (WHO, 2001, Table 5.1, p. 114). In all parts
of the world these recommendations included: educating the public about mental
health and mental illness through campaigns to reduce stigma and discrimination
and foster more positive attitudes towards people with mental illnesses; linking
with partners in non-medical sectors to promote mental health through school
and work place programmes; and monitoring mental health in the community. A
number of other measures were also recommended that focused on developing
national mental health services in countries where these were very rudimentary, or,
where services were already in place, improving the level and range of community
based provision and the effectiveness and equity of services for all users.

These recommendations were subsequently taken up, for example in WHO
policy statements such as the *Mental Health Declaration for Europe* (WHO, 2005)
agreed by the Health Ministers of countries in the European Region of WHO (see
Box 8.1). Their statement was further reinforced by a publication summarising
the agenda they put forward (WHO European Region, 2005). In the USA, the
President's New Freedom Commission on Mental Health (Commission on Mental
Health, 2003) also addressed this agenda. Points from the declaration in Box
8.1 also emphasise the need to address aspects of social environments which
disadvantage people with mental illness and to take into consideration provision

Box 8.1   The WHO (2005) Mental Health Declaration for Europe

The World Health Organization (2005) published a *Mental Health Declaration for Europe,* agreed by the Ministers for Health in European countries which refers to the recommendations relating to mental health in the WHO *World Health Report 2001* and approved certain developments in mental health policy and services of countries across the European Region which they describe as follows:

"...striving to achieve social inclusion and equity, taking a comprehensive view of the balance between the needs and benefits of diverse mental health activities aimed at the population at a whole, groups at risk and people with mental health problems. Services are being provided in a wide range of community-based settings..." (WHO European Ministerial Conference on Mental Health, 2005: 2)

The signatories to the declaration further stated that they "...welcome the fact that policy and practice on mental health now cover:
...the promotion of well-being;
...the tackling of stigma, discrimination and social exclusion;
...the prevention of mental health problems;
...care for people with mental health problems, providing comprehensive and effective services and interventions, offering service users and carers involvement and choice;
...the recovery and inclusion into society of those who have experienced serious mental health problems".

However, they also expressed the view that "...the main priorities for the next decade are to:
...foster awareness of the importance of mental wellbeing;
...collectively tackle stigma, discrimination and inequality, and empower and support people with mental health problems and their families to be actively engaged in this process;
..design and implement comprehensive integrated and efficient mental health systems that cover promotion, prevention, treatment and rehabilitation, care and recovery;
...address the need for a competent workforce, effective in all these areas;
...recognise the experience and knowledge of service users and carers as an important basis for planning and developing mental health services." (WHO European Ministerial Conference on Mental Health 2005: 2–3).
To achieve these priorities requires action involving communities to address the wider determinants of health.

The declaration (WHO, 2005: 3–4) therefore also calls for actions in areas that include the following:

..."promote the mental well-being of the population as a whole by measures that aim to create awareness and positive change for individuals and families, communities and civil society, educational and working environments, and governments and national agencies"
..."consider the potential impact of all public policies on mental health, with particular attention to vulnerable groups, demonstrating the centrality of mental health in building a healthy, inclusive and productive society"
..."assess the mental health status and needs of the population, specific groups and individuals in a manner that allows comparison nationally and internationally...initiate research and support evaluation and dissemination" of the actions proposed."

These objectives highlight the relevance of the perspectives considered in this book because they focus attention on mental health at the population level, call for action in a variety of different socio-spatial settings, require attention to a range of processes in the wider environment that influence material and social experiences, and suggest that comparisons should be drawn between different contexts at various geographical scales in order to assess and investigate inequalities.

In addition, a larger number of other actions were declared as important to improve the clinical and social treatment of people with mental illness. These included the need to:
"tackle stigma and discrimination...empower people at risk of suffering from mental health problems and disabilities to participate fully and equally in society...offer people with severe mental health problems effective and comprehensive care and treatment in a range of settings and in a manner that respects their personal preferences and protects them from neglect and abuse..." (WHO European Ministerial Conference on Mental Health, 2005: 3–4)

Research which demonstrates how people with mental illness experience different social environments and the social relationships which affect their social inclusion, empowerment and autonomy is therefore relevant to this policy agenda.

of care and support through a complex network of care settings, which may be variably experienced by different individuals, and which are important for social integration, as well as medical treatment of people with mental illness. The stress placed on education and work environments reflects the relevance of health determinants in a range of settings. It can therefore be argued that several aspects of the population mental health agenda and studies of the ways that individuals' mental health is influenced by their material and social context show a close fit with the research agenda for geographies of mental health. The international policy debate surrounding human mental health and its sustainability in the face of contemporary risks to health, therefore places considerable emphasis on population mental health. The following discussion considers how the approaches discussed in this book can inform this debate.

Measuring and monitoring population mental health at the national and local scales

One aspect of this growing concentration on promotion of mental health at the population level has been development of methods to compare levels of mental health and track progress in mental health gain, that are applicable to the population as a whole, and not solely based on information about people with mental illnesses. Geographical approaches make an important contribution to this perspective on mental health, partly because they gather and analyse information on the health of groups of people populating particular geographical spaces and on factors associated with geographical variability in population health.

As noted in Chapter 1, community surveys can be used to assess general mental health independently of psychiatric service use, using questionnaires to assess self-reported experience of signs and symptoms of distress. Several of the studies reviewed in earlier chapters, concerned with environmental factors associated with mental illness, have made use of these kinds of measure of mental disorders. National sample surveys (such as the *Health Survey for England*, the *Community Health Survey* in Canada, the *Health Interview Survey* in the USA) all employ these kinds of indicator to assess levels of psychological distress and mental illness in their populations and are repeated periodically so that it is possible to assess trends.

However, in order to address a health promotion agenda, it is necessary to go beyond assessment of population mental health in terms of psychiatric disorders or psychological morbidity. It is also important to consider a wider range of ideas about positive mental health and wellbeing, taking into account that mental health is socially as well as clinically constructed and can be defined in different ways. In Chapter 1 some ideas about conceptualising and measuring wellbeing and quality of life were introduced. These approaches are relevant to efforts to measure wellbeing in order to inform policies at the population level, although they still tend to be overly focused on negative aspects of mental state and do not

a) Proposed constructs for the Indicators

High level constructs		
Positive mental health		Mental health problems
Contextual constructs		
Individual	Community	Structural/policy
Emotional intelligence	Participation	Violence
Spirituality	Social networks	Physical environment
Learning and development	Social support	Working life
Healthy living	Trust	Stigma/discrimination
Physical health	Safety	Debt/financial security
		Social inclusion
		Equality

b) Sample questions from the Warwick-Edinburgh Mental Well-being scale (WEMWBS)

Below are some statements about feelings and thoughts.
Please tick the response that best describes our experience of each over the last two weeks.
Response scale: None of the time / Rarely / Some of the time / Often / All of the time

I've been feeling optimistic about the future
I've been feeling useful
I've been feeling relaxed
I've been feeling interested in other people

Figure 8.1   Example of an approach to assessment of mental wellbeing used by Scottish Government

Source: http://www.healthscotland.com/documents/1467.aspx.

always measure variability in positive mental states with much precision. Some questionnaires are also too long and detailed for use in large population surveys covering a number of aspects of health. Therefore further research is aiming to develop measures which will be more closely aligned with specific national and local policy goals and will make it possible to measure different degrees of positive psychological state, using fairly simple self report methods. To take one example, we can consider the *Well Scotland* initiative promoted by Scottish Government, which has identified some key factors likely to contribute to positive mental health (Figure 8.1a) and is working to collect indicators which will assess these. *Well Scotland* has adopted a variant of the *Affectometer2* questionnaire developed in New Zealand by Kammann and Flett (1983) and tested in the Scottish context by Tennant, Joseph et al. (2007). They are also using the a measure of positive wellbeing developed by Warwick and Edinburgh Universities illustrated in Figure

8.1b (Tennant, Hillier et al., 2007; Stewart-Brown et al., 2009). Similar kinds of measure are being explored in England (e.g. DEPFRA, 2008). These measures can be helpful when it is possible to survey the population to collect detailed information on self reported wellbeing. However, there is currently a gap in research on how to formulate small area indicators relating to psychological wellbeing that could be updated regularly from routine sources without using special population surveys, and would be applicable to local areas across the country.

Quite apart from problems of measurement, the question of how to conceptualise positive aspects of psychological wellbeing also presents a challenging research agenda. As discussed in Chapter 1, Carlisle and Hanlon (2007) and Hanlon and Carlisle (2008) have drawn attention to theories and empirical research in positive psychology and philosophy, which suggest that positive psychological wellbeing is not only a question of short term happiness, but may also be bound up with other aspects of a 'flourishing life'. These may include having a sense of resilience in face of adversity and being able to achieve goals that are challenging, but well adjusted to what is possible in life. Also reviewed in Chapter 1 was work by Saxena and colleagues (2001) suggesting that cross cultural comparisons of these dimensions of a 'good' and 'satisfying' life may be quite problematic, since valued roles for the individual in society are variable from one culture to another. Research agendas in positive psychology are concerned with the relationships of positive emotions, positive character traits and social and institutional structures which affect psychological state. These topics resonate with research in emotional geographies, especially the themes considered in Chapter 4 and 6 of this book.

Social and territorial equity in health care provision

Aims to prioritise population mental health, discussed above, call for greater efforts internationally to improve equity in access to mental health services, including community based services. The discussion in Chapter 7 of this book drew on post-asylum geographies of care for mental illness and psychological distress and showed that there is a matrix of care settings important for treatment of mental illness and alleviation of distress, partly through the delivery of medical care, but also through social integration of people with mental illness. This matrix of care raises questions of equity in mental health service provision and the challenge of matching mental health service resources to variations in 'need' in populations of different geographical areas.

This is an important issue; many countries in the world aim through their national health service systems to provide access to psychiatric care in a way that ensures that psychiatric services are available to those who need to use them and that national resources are used effectively and efficiently to treat mental illness. Davies (1968) coined the term 'territorial justice' to express the idea that national administrations should try to ensure a fair distribution of resources for health and social care for different geographical areas (territories) within their jurisdiction. Similar arguments are put forward from a geographical perspective by Smith (1977). The argument rests on two ideas. The first is a view originating in philosophical theories of social justice (e.g. Rawls, 1972) that it should be possible to define a 'normative' principle of what constitutes a just distribution of social resources. The second is that some steps can be made towards operationalising this principle of a just distribution, through service planning and resource distribution, even if present knowledge and techniques for this are imperfect. Most attempts to define such 'normative' definitions of 'need' and a fair distribution of resources include the principle that care and treatment for mental illness should be allocated in proportion to the 'needs' of people with psychiatric disorders or other mental health problems that may benefit from receipt of services. While arguments about territorial justice have mainly been put forward in relation to discussion of fair distributions of resources within particular countries, similar arguments can be applied to international distributions, among countries, of mental health care resources in relation to population need. Discussions about equity in mental health care often draw on geographical arguments and methods, which are of particular interest in this book.

At the international level, as noted in Chapter 1, it is difficult to accurately compare the mental health of the population in different countries in order to assess relative needs. Governments in poor countries may prioritise physical health problems over mental illnesses because they are relatively important causes of mortality and morbidity and are less well controlled than in wealthy countries. However, this does not necessarily mean that in low income countries levels of mental illness requiring psychiatric health care will be relatively much less than in richer countries. Also, given the strong links between social disadvantage and mental illness discussed in earlier chapters it might be argued that in all countries the poorest populations are likely to have proportionately greatest need of collectively provided mental health care. Thus at the international level the question of territorial justice may be posed in terms of whether countries in different parts of the world provide relatively similar levels of mental health services to their populations generally and whether provision ensures access to psychiatric care for the poorest populations in particular.

An analysis providing some indications on these points has been published in the World Health Organization *Mental Health Atlas* (WHO, 2005). This is really an atlas of mental health *care* as it presents analyses of data on provision of psychiatric health services in over 180 countries in all regions of the world. The atlas includes a report on each country and also summarises some aspects of the pattern of variation among countries that vary by income group and by world region. For example, Figure 8.2a shows that in all of the 'middle to high' and 'high' income countries, the primary source of funding for mental health care is tax-based state funding or social insurance, but in low income countries, only 40 per cent use this as the main way to fund mental health care, and many rely on out of pocket payments by people who need to use health services. This implies that inequality of access to mental health care which disadvantages the poorest populations is most likely to occur in

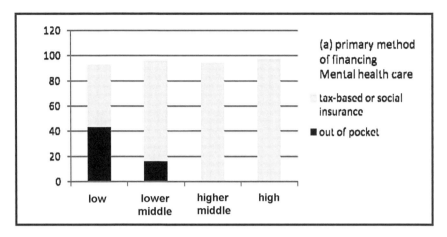

8.2a

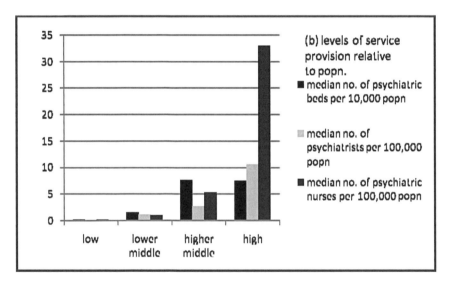

8.2b

Figure 8.2   Information from the WHO (2005) Mental Health Atlas showing differences in provision of mental health care among countries grouped by national income

Note: Based on data from WHO (2005) *Mental Health Atlas 2005 Revised Edition.* Retrieved from http://www.whot.int/mental_health/evidence/mhatlas05/en/index.html, Tables 35, 41, 48, 52.

low income countries. Also the level of provision of psychiatric beds and psychiatric doctors and nurses relative to the total size of the national population is much higher in high income countries than in low income countries (see Figure 8.2b). Saxena et al. (2006: 179) argue that overall, this atlas indicates that "global mental health resources remain low and grossly inadequate to respond to the high level of need" and that "Imbalances across income groups of countries remain largely the same". They suggest that "more resources are needed to meet the mental health needs of populations around the world...especially in low- and middle-income countries". This argument might be considered contentious, given the variable quality of the available statistical evidence, cultural differences in views about the nature of mental illness, and the best way to provide care for mental illness, and varying national priorities for health care funding. Nevertheless, the *Atlas* does seem to suggest that low income countries have very little resource to support comprehensive mental health services and it seems likely that in these countries a relatively large proportion of people with serious mental illnesses, especially the poorest people, will receive little or no psychiatric care.

At the intra-national level, not all countries consider equity and cost effectiveness of health service provision in the same way, and the approaches used depend on national ideologies of care and the ways that service provision is organised and financed. Other texts on health geography have discussed in some detail these ideological and organisational differences (e.g. see Jones and Moon, 1987; Curtis and Taket, 1996, Chapter 5; Mohan, 2002). In spite of these differences, many of the countries with well established national health service consider that there should be collective action by the state to ensure that psychiatric services are made available to all members of society that need psychiatric treatment, even to those individuals who cannot afford to pay the whole cost of these services themselves.

For example, health services in the USA are funded to a large extent through private insurance. This country has historically remained ideologically opposed to the idea that comprehensive medical care, as a basic human right, should be made universally and equally accessible to all citizens at the expense of the taxpayer (although health care legislation passed in 2010 may reflect some change of view). However, a large proportion of psychiatric care used by American Citizens is financed through public funding schemes such as 'Medicaid' or, for older people, 'Medicare'. These schemes are designed to provide a 'safety net' to ensure that basic care is available to people without health insurance. This means that, in practice, there is a concern to understand the factors that may drive demand for psychiatric care and to ensure that public funds are used appropriately and efficiently in all parts of the country to meet these needs.

In Britain, the majority of the population use the National Health Service (NHS) for most of their health care needs. The NHS is funded through a central state health care budget, and is based on the founding principle that access to health care should depend on the nature and urgency of medical need and should not depend on financial means or area of residence. Therefore, in the British case the aim is to make territorial distribution *equitable* as well as cost-efficient. This means

Type of health system	Example	Use of information on estimates of 'needs' for psychiatric care in the population of local areas
Independent insurance (with significant numbers lacking insurance); state funded 'safety net' schemes to cover costs of basic services for some people with long term illnesses who do not have insurance	USA ('Medicaid' or 'medicare' and private insurance)	To assess efficiency in provision of services and expenditure by 'safety net' funding and identify local anomalies in state funding of psychiatric services
Comprehensive, collectively financed National Health Services, funded mainly through taxation. Some private health insurance for a minority of the population	England (National Health Service)	Determine how the central budget for health services provision should be distributed to local areas and check that spending in local areas is proportional to estimated population needs
Comprehensive social insurance schemes providing 'universal' cover for health care costs to all citizens, some of it on the basis of reimbursement of a proportion of total costs	France (universal insurance through various insurance funds)	To assess whether health service capital infrastructure and human resources are fairly distributed in the national space and to monitor local expenditures relative to needs

Figure 8.3   Different types of national health service systems and their use for local area estimates of need to use services

that resources made available for psychiatric services at the local level should be proportional to a 'normative' definition of relative need to use services in the population. NHS managers are therefore concerned to determine what factors in the population might be used to predict variations in need for psychiatric care and compare them with actual levels of service spending and provision.

Similar arguments apply also in countries with social insurance systems for health care (including many countries in continental Europe) which are designed to ensure universal coverage, especially as these nations usually aim to provide health services equitably relative to health care needs in different parts of the national space. Although their funding mechanisms may make it more difficult to actively direct funds to local areas in proportion to 'normative' 'need' measures, their planning systems often monitor variation in capital infrastructure and human resources to make sure access to psychiatric services is fairly distributed across all areas and ensure that it is proportional to local population needs.

Figure 8.3 summarises these arguments, showing that, in countries with health systems as diverse as those of the USA, France and Britain, there is government interest in the question of how to define and measure local indicators of 'need' for health services at the population level and to compare these with information access to, and use of health services. However, defining need and use of services at the local scale, is a difficult problem that is not fully resolved, even in countries where considerable effort and has been devoted to the matter (Box 8.2 and Figure 8.4). (For a more general and detailed discussion of these issues from a geographical perspective see Curtis, 2004). Various analyses have examined the statistical association between socio-demographic variables and psychiatric morbidity (e.g. Kessler et al., 2004; Congdon, 2006) or mental service use or mental health care expenditures for small areas across a country or region as a whole (Jarman et al. 1992, Glover et al., 1998, 2004; Smith et al., 1996, McCrone et al. 2006; Curtis, Congdon et al. 2009). Other studies, cited in Chapter 5, have examined these links for small areas at the level of particular cities. These studies, which are all from 'western' countries with well established health information systems, mainly focus on admissions to psychiatric hospital and suggest that in addition to overall population size and age and sex differences in hospital admission are associated with variables reflecting levels of poverty and material deprivation, indicators of social isolation and fragmentation, and, in some studies, concentration of certain ethnic groups. Since these indicators reflect population characteristics and processes which health services cannot alter, they might be viewed as proxy measures for varying 'need' for psychiatric care. A fair and efficient distribution of services will tend to distribute health care resources in proportion to these population differences. In Britain information from studies of this type are used by central government to inform resource allocation for NHS psychiatric service provision to different parts of the country (Department of Health 2005, 2008). However, the deployment of these resources in different ways to provide various types of NHS psychiatric care is decided locally, not centrally.

Efforts continue to produce more sophisticated statistical models of varying patterns of population 'need' and psychiatric service use. For example, Curtis, Congdon et al. (2009) attempted to tackle some of the deficiencies of other models in a study of psychiatric service use in counties of New York State. They aimed to take account of the complex relationships between use of inpatient and ambulatory psychiatric services, as well as characteristics of the local population. Their model showed the relationship of population variables with both ambulatory service use and hospitalisation rates across counties. Their method also combined and weighted population variables in the 'need index' in a way that (a) reflected geographical variations in service use rates in both hospital and ambulatory sectors (b) controlled for service configuration and access as well as (c) allowing for spatial autocorrelation in the population 'need' index. Their study therefore moves a step closer to being able to represent the complex local variability in use of different services in the matrix of care.

Box 8.2   Different approaches to estimating population needs for mental health services at the small area scale

A number of studies have used statistical models to examine how geographical variation in population 'need' and psychiatric service provision vary together across local areas within countries. Broadly speaking, two approaches are common, based on estimating local levels of *psychiatric morbidity* or on estimating likely levels of *psychiatric service use* (Also summarised in Figure 8.4).

One approach to defining local 'need' is to try to estimate psychiatric morbidity in the population at the level of local areas across the country and use this as a measure of relative population 'need' for services. A strong positive correlation between the local level of provision and the local level of psychiatric morbidity would usually be interpreted as indicating an 'equitable' and 'efficient' pattern of provision. This is challenging because it is difficult to assess mental health status of the population at local level in all parts of a country, especially if one is looking for a measure of need which is independent of service use. Some research uses methods of 'synthetic estimation' of *prevalence of psychiatric illness*, which predict local levels of psychiatric disorder. Local levels of morbidity are predicted from ecological data in the form of socio-demographic variables for small areas (usually small area census statistics, available for all small areas around the country). To estimate morbidity these variables are selected and weighted on the basis of associations with psychiatric illness which are observed in national surveys (e.g. Kessler et al., 2004; Congdon, 2006).

Especially in the UK which has very comprehensive routine data on psychiatric service use through the National Health Service, an alternative approach to investigating how population variables relate to 'need' for psychiatric services has also been quite widely applied. This also uses ecological information for small areas on locally varying socio-demographic variables, based on routinely available sources such as the population census. The variables of interest are selected and weighted according to their association with psychiatric health *service use* in the same small areas. They are then combined into 'population need indices' that can be interpreted as 'proxy' measures of a 'normative' population 'need' for psychiatric care. An assessment of equity and efficiency can then be made by comparing the actual rates of use of services in particular local areas with the level that would be expected if the local area had rates of use that were consistent with the 'typical' pattern, averaged for other areas with similar populations across the country.

Examples include (Glover et al., 1998, 2004; Smith et al., 1996) who have selected and weighted demographic variables for small areas in England according to their relationship with local rates of use of psychiatric hospital beds, and controlling for geographical proximity to psychiatric hospitals. Indicators similar to the ones they have derived are used to help determine the distribution of funds for psychiatric care in the English NHS, for example (Department of Health, 2005). McCrone et al. (2006) report a related approach, also using local socio-demographic variables, but weighting them by their relationship with social services expenditure for psychiatric care.

Statistical models of the ecological variability of psychiatric service use in small areas have potential to provide useful information for service planning, but they also have a number of limitations (See Figure 8.4). These include technical problems: lack of information about the full range of psychiatric services in the 'matrix of psychiatric care (discussed in Chapter 7), comprising community based services as well as hospitals; failure to take account of spatial proximity to services which is often positively associated with use of services; failure to consider the possibility that clusters of small areas may share certain patterns, generating problems of spatial autocorrelation in the models.

There are also more basic theoretical issues in terms of how to translate the findings from these analyses into prescriptions for a just and efficient distribution of services. The first method described above, estimating illness prevalence, has the advantage that it might be assumed to be 'independent' of current practice in service provision. However, this assumption does not take into account that the distribution of the population with psychiatric illness may partly depend on current service provision (e.g. see the discussion of 'drift' effects in Chapter 7). Furthermore, it is not always obvious how to translate information on estimated levels of psychiatric illness into predictions of use of services. The second method, estimating service use, is based on more direct information about population factors that predict use of psychiatric services, but is criticised because, in using information on the existing geography of service use, it may tend to reproduce the existing pattern of service provision nationally. This will not necessarily be the best solution in terms of territorial justice or service efficiency. Another problem for both methods is that models like these of relationships applying to all small areas in the country may not be able to take into account specific local factors which are important for mental health service needs in particular areas.

Strategy for estimating local population 'need' for psychiatric services	Approach	Examples	Normative definition population 'need'	Strengths	Weaknesses
Synthetic estimation – estimate the relative numbers of people with psychiatric disorders in local populations	Use national sample survey data to establish the association between demographic variables and service use for survey respondents. Use the results to select and weight variables which can be used to estimate local 'need' for psychiatric illness	Kessler et al. (2004) in USA; Congdon (2006) in England	Relative size of estimated numbers of people in the local area with psychiatric disorders	A 'pure' 'need' measure that is relatively independent of service use.	May be difficult to make accurate estimates of morbidity that are sufficiently discriminating of different levels of morbidity. Population surveys are costly so may be difficult to keep the estimates up to date.
Demographic factors known to be ecologically associated with psychiatric service use at the local level	Use small area information on psychiatric service use as dependent variable in a regression analysis that predicts service use from demographic variables. Use the regression a coefficients to select and weight demographic variables to include in local 'need' indices	Smith et al. (1996); Glover et al. (1998, 2004); McCrone et al. (2004)	Relative levels of service spending/provision in each small area should be proportional to the 'need' index. Small areas where relative provision/spending is not well predicted by the index are considered not to correspond to normative need	Variation in psychiatric service use can be estimated from routine data (at least in comprehensive systems like the English NHS) the needs index can be updated regularly. The weighting of the needs index reflects the average level of psychiatric service provision and spending that is generally considered appropriate in areas around the county	If the pattern of provision that is typical across the country is not the most efficient or effective one, this system will tend to perpetuate the inefficiency. There is no measure of 'need' independent of current service use. Method will not work if comprehensive information on health service use is unavailable at small area level.

Figure 8.4   Comparison of two ways to estimate 'needs' for psychiatric services in the population of local areas

In future we may expect to see further advances in modelling of both hospital and community services, and it would be helpful to be able to model time trends in service use in order to assess change in the relationships between population variables and psychiatric service use. This information could inform policy by demonstrating whether patterns of provision relative to population 'needs' for services are shifting towards more equitable and efficient distributions. There is also potential to do more to model these relationships at different geographical scales, to see whether there are clusters of small areas in 'regions' which seem to have distinctive patterns of psychiatric service provision, unlike those in the country as a whole. This would raise questions over whether single national models of 'ideal' levels of provision relative to need are appropriate for all parts of the national space.

Organising mental health care to suit the local context: examples from rural settings

Although statistical modelling can usefully inform planning for the general distribution of resources to different areas, these need to be used variably at the local level to provide the types of services that are suited to local conditions. The question of whether local service provision matches population 'needs' can be usefully informed by more intensive, qualitative studies of perception of services in particular places, especially in areas which have distinctive attributes. Here the aim is to explore the complex relationships between the people using services, their geographical context and the organisation of local services in order to judge how far services are accessible and well suited to local population needs. For example, Parr, Philo, C. (2003: 471) note a 'move away from research...that constructs mapped spatial distributions of...health care resources and access patterns, towards theorising care and caring relations in ways that highlight diverse political and economic processes involved in care restructuring'.

One group of studies providing interesting examples of research in this field have explored access to health services for rural populations in Scotland. As noted in earlier chapters, rural areas in Britain rank relatively poorly on some mental health indicators (for example, suicide rates are relatively high in remote rural areas compared with moderated urbanised and semi-rural areas) so that mental health care provision to meet local needs in rural areas is an important issue. Philo, Parr and Burns (2005) comment on the lack of attention to experiences of mental health service users in rural areas, compared with the more extensive research conducted in urban settings. Authors such as Milligan (1999), Parr et al. (2004), Philo et al. (2005) and Bondi (2009) describe the experiences of people living in rural parts of Scotland who suffer from mental illness or psychological distress or are providing mental health services. In various ways, these accounts all emphasise experiences of inclusion and exclusion for mentally ill people in rural communities which combine 'contradictory' tendencies in these rural societies. On one hand, sparsely populated rural communities are made up of people who are often spatially distant from each other. On the other hand, people are used to travelling some distance for services or social gatherings and some members of rural communities feel socially close to their neighbours even though they are geographically distant. Most people are known to each other in small communities and this may create an atmosphere of intimacy and acceptance which encourages practices of mutual support and caring.

However, the experience of social rejection or stigma, which is common for people with mental illnesses in many settings, can feel more pronounced and painful in a small, inward looking community. These studies from rural Scotland show that for individuals who do not benefit from the 'social proximity' that binds rural communities together, the sense of social isolation and exclusion can be more extreme than in urban settings where anonymous interactions with other people are more frequent. Philo et al. (2005: 405) report numerous examples of exclusion including extreme forms such as raising a petition against the presence of a mentally ill person in the community, pointing out and (children) shouting abuse, turning away from the person (the 'body swerve'), refusing to acknowledge people. They also note that rather different 'expectations' and 'tolerances' were reported for behaviour of 'locals' compared with 'incomers', although the results could vary. Incomers might be rejected if their behaviour seemed strange, or might be more tolerated because their origins meant that they would be expected to behave 'unusually'.

Because there is little chance of anonymity in rural settings, confidentiality in the treatment setting can be problematic. Therapists' attempts to treat their clients confidentially may be undermined because they are likely to meet them outside the consultation room in other roles. Bondi (2009) describes how a psychotherapist chose to set up a consultation space in a hotel, partly because, being a place used by a variety of people from within and outside the community, clients could visit the building without attracting attention. Parr and Philo (2003) cite the example of a community psychiatric nurse who parked her car a mile away from the isolated home of a user to try to avoid drawing attention to the fact that the person was using psychiatric care.

Parr and Philo (2003) also discuss in more general terms how health service provision in rural Scotland is complicated by the relative lack of specialised resources, not only in terms of facilities, but also in human resources, so that, for example, general practitioners and home nurses may find they have to deal with more complex mental health needs than would be typical in urban areas with more specialist staff and facilities. Some services have to be provided in patients' homes or other settings that are not always well designed for the purpose. They comment on the significance, in these kinds of settings, of community action and organisation around declining services and problems of access to services that are only available at a distance. They also note the relatively high degree of dependency on social support as a 'safety net' providing informal care in place of professional services.

While the Scottish Highlands and Islands region has received particularly intense attention on the part of British health geographers interested in mental health issues, this is not, of course the only part of the world where these issues of rural mental health care apply. Problems for otherwise developed national health systems in ensuring adequate provision of mental health care in remote areas are discussed in studies from other countries which have more extensive and more isolated rural spaces. (e.g. Moore and Nelson, 1998; Boydell et al., 2006; Ryan-Nicolls et al., 2007; Collins et al., 2008, 2009; Taylor et al., 2009). Several of these studies draw conclusions similar to findings in the geographical work in Scotland considered here concerning the challenges as well as the advantages of rural socio-geographical settings for the provision of mental health care. These studies underline the importance of variation in social and material environments for individual experiences of mental illness and mental health care, and they illustrate the interactions between individual and environmental conditions that have been discussed throughout this book.

Provision of psychiatric care in very remote areas also raises some interesting questions about the scope for delivery of services via telecommunications systems. The now quite widespread experience of tele-psychotherapeutic services for rural populations has been reviewed, for example, by Picot (1998), Janca (2000) and Norman (2006) and it raises possibilities for rethinking models of health care to suit rural populations. The experience of health and health care in virtual settings, through tele-psychiatry, also poses theoretical questions for health geography concerning the importance of 'imagined', virtual spaces for health, which Cutchin (2002) has suggested constitute a 'new frontier' for this field of research. It also involves a recasting of the ways that we think about key roles and relationships in health care provision (Nicolini, 2006) and the relevance of geography in arguments about social justice (Bauer, 2003).

Strategies for mental health promotion: acting on the wider determinants of health

In order to promote good mental health and prevent illness in the population, there is a much wider agenda to be addressed, extending beyond issues of mental health services to concerns about general material, social and symbolic environmental factors that have been reviewed in this book.

The Health Ministers of European countries have declared (WHO Regional Office for Europe, 2005: p. 2) that mental health promotion at the population level was an important part of their strategy, to "enhance mental well-being for all". They state that, "Public mental health reinforces lifestyles conducive to mental wellbeing. Mental health promotion needs to target the whole population, including people with mental health problems and their carers." They also commented on the importance of reducing stress which is harmful to mental health. They argued that: "The social causes of mental health are manifold, ranging from individual causes of distress to issues that affect a whole community or society. They can be induced or reinforced in many different settings, including the home, educational facilities, the workplace and institutions". They also highlight that certain marginalised or vulnerable groups may be at particular risk of stress that affects mental health.

Associated with these kinds of objectives for public mental health there has been a considerable growth of interest in how societies and individuals might manage aspects of their environment better in order to promote good mental health. In European and Australasian countries and in parts of North America, there has been a particularly strong debate in the field of public health which aims to make various aspects of social decision making and human action more sensitive to the implications for health, including mental health. Increasing emphasis is placed on 'Health Impact Assessment' of policies and actions in all areas of public sector activity, and to an extent in the private sector as well (see for example WHO European Centre for Health Policy (1999), Douglas et al., 2001; Mindell, Ison and Joffe, 2003, Kemm et al., 2004). Mental Health/Wellbeing Impact Assessment focuses particularly on the mental heath and psychological outcomes of policies and actions (Coggins et al., 2007). Health Impact assessment can be carried out in many different ways, but is commonly designed as a process to assess prospectively, before implementing a policy or action, what the impacts might be on public health or on health inequalities. Typically it is applied to plans and interventions in sectors outside medicine. The process draws on existing evidence about the health impacts of other, similar interventions in the past, and it is also supported by evidence of the type reviewed in this book. Health impact assessment also often draws on 'civic intelligence' and expert opinion through consultations with key informants and stakeholders in the scheme that is being proposed.

A further sign of the increasing focus on mental public health has been a growth in attention in public health publications to discussions about the ways that societies can take initiatives to make changes to physical and social environment at the community level in ways which will benefit health. For example a special issue of the *American Journal of Public Health* in 2003 (No. 93, 6) was devoted to the question of how changes to the built environment could contributes to health promotion. Papers in this collection include one by Frumkin (2003) who focuses on the idea that a positive sense of place may be important for psychological and physical health. He argues that promising strategies may include changes to the built environment (contact with natural elements, design of buildings and public spaces, and the general urban form including transport networks and juxtaposition of different land uses). Similar arguments are put forward in companion papers by Kaplan and Kaplan (2003) who argue that people become more 'reasonable' in their behaviour and relationships with others and are more satisfied with their environment when it includes the kinds of material aspects that were reviewed in Chapter 2 of this book. The collection also features a discussion by Saegert (2003) examining the evidence from US studies of housing modification in relation to public health outcomes and a consideration of how the indoor environment relates to health by Samet (2003).

Twiss (2003) and Milligan, Gatrell and Bingley (2004) report on the potential of community gardens as examples of spaces which promote wellbeing through therapeutic contact with nature and with people. Similarly, in the UK, a recent publication by the Young Foundation (Hothi et al., 2009) puts forward arguments about how changes to the social environment of neighbourhoods that encourage 'neighbourliness' and 'empowerment' are likely to increase wellbeing for all members of local communities. This draws on the findings of research similar to that reviewed in Chapter 4 of this book concerned with the benefits of social inclusion and social capital for psychological health. Also the National Institute for Clinical Excellence published a report on interventions that have been shown to be effective in improving mental health through action in primary health care, but also outside the medical sector in areas such as the workplace (Taylor et al., 2007).

These reports on the potential to make changes in social and material environments that will benefit health often stress the theoretical potential of such interventions while also emphasising that there is limited empirical 'evidence' that making changes of this type will actually produce health changes. For example, among the papers just cited, Saegert (2003); Frumkin (2003) and Taylor et al. (2007) are all quite critical of the scientific strength of the research evidence in this respect. (See also the discussion by Mindell et al., 2004 concerning how to enhance the evidence base for health impact assessment.) This raises an interesting conundrum for public mental health, since on one hand there is strong and growing commitment to the idea that it is necessary to act on the material, social and economic determinants of health in order to achieve real change in population health, but on the other hand, public health specialists seek an evidence based approach that calls for research findings that often do not exist.

Mental health and sustainable societies

The research discussed in this book supports this public mental health agenda through its emphasis on aspects of environment that benefit mental health including:

- *Material* features (e.g. natural landscapes that are beneficial to health and build forms that offer 'architectures of wellbeing');
- *Social attributes* (e.g. greater levels of social inclusion, healthy forms of social cohesion);
- *Symbolic dimensions* (e.g. appreciation of symbolic aspects of environment that are variable according to one's frame of reference and which support positive senses of identity, self esteem, belonging and security and opportunities for spiritual growth, creativity and fulfilment).

However, the research reviewed here also repeatedly emphasises that the *interactions* between these different aspects of environment are crucial for the ways they are associated with mental health, and that furthermore the way that individuals respond to environments is variable.

In this book it has been necessary to be very selective, but even from the material included here it is clear that it is probably necessary to take a very broadly based strategy for mental public health that extends beyond specific 'public health' interventions. In order to do so, it will be necessary to consider quite complex models of the combinations of processes through which mental health relates to its wider determinants. Furthermore, the complexity of these interrelationships suggest that it will be extremely challenging, if not impossible given current scientific methods, to generate the kind of scientific evidence base that some commentators argue to be necessary to 'prove' that benefits mental health and wellbeing can be clearly attributed to changes in specific aspects of environments.

One example of this more complex way of thinking is an argument put forward by Aneshensel (2005) in support of the 'Social Consequences' model which views socially unequal health outcomes as a 'by-product' of widely accepted social structures. These processes that normally operate in ways that must be expected to produce social differences in many aspects of life. Aneshensel points out that this perspective is contrary to the idea that health inequalities are due to a 'malfunction' of 'normal' social processes, which can be corrected through specific interventions, without changing the fundamental structures. The Social Consequences model is quite challenging because it interprets differences of mental health and illness at the population level as closely associated with the social structures and processes in each society, suggesting that 'systemic' changes would be needed to tackle these health differences.

Several of the themes highlighted in this book have also raised important issues concerning human mental health as an aspect of the more general *sustainability* of societies around the world. We noted in earlier chapters that contemporary human societies face a range of challenges in the material and social environment which Beck (1999) discusses in terms of the 'world risk society'. These global risks include the possibility of 'disaster' events and more persistent risks in the material and social environment that are known to be associated with mental illness and distress, as discussed in earlier chapters of this book. For example, global warming brings increased risks of extreme weather events and associated environmental hazards such as floods, storm damage and drought. Geological events including earthquakes, landslides and volcanic eruptions, are increasingly likely to threaten large populations because major cities have grown up in the areas at risk and, furthermore, human societies are increasingly dependent on extensive and complex infrastructures such as roads, pipelines, hydrological dams, etc, which may be damaged. Disasters including major industrial pollution episodes, pandemics of disease, global economic recession and terrorist attacks are also commonly discussed. Apparently less dramatic, but perhaps of even greater significance because of the large numbers of people affected over extended periods of time, are challenges of persistent environmental contamination and threats to biodiversity, as well as poverty and social disadvantage affecting large proportions of the world's human population, with major inequalities between rich and poor people persisting and increasing within and between countries.

These changes therefore raise questions for human society globally, concerning its fragility and lack of sustainability, generating a general sense of anxiety which Beck (1992, 1999) has conceptualised as a defining feature of present day 'risk society'. Thus one can argue that concern for mental health of the population as a whole is one of the considerations driving efforts to ensure greater sustainability of human societies around the world. Particular emphasis needs to be placed on public mental health perspectives that are sensitive to the links between mental health at the population level and sustainable physical and socio-economic environments.

Beck's analysis is also useful to this discussion in view of his arguments about non-linear knowledge and our ability to take an 'evidence based' approach to risk (see Chapter 1 of this book). I have argued elsewhere, for example (Curtis, 2008:), that "the scenario for health impact assessment of policies and actions affecting the wider determinants of health seems very similar to that described by Beck as typical of 'non-linear knowledge' in 'world risk societies'. Especially when we are dealing with complex social or economic interventions, the factors influencing risk are not sufficiently well known to be able to attribute risks or to assess them, especially in quantitative terms, so the risks are often incalculable. The available research does not provide a complete picture of what we need to know (a large degree of unawareness) [and there are] multiple, competing opinions about what counts as 'reliable' evidence".

This scenario seems to call for acknowledgement of relatively complex problems we face in dealing with the wider determinants of mental health, and an acceptance that responsibility for addressing these is widely distributed in society. Conceptual frameworks concerning the challenge of sustainability of human life may be relevant here because these models emphasise the need to extend efforts to promote sustainability quite widely and to recognise the interconnected nature of ecological systems.

For example, Veil et al. (1992: 409) call for action on a 'healthy environment', "conducive to physical and mental health", which they argue to be a human right for all people. To achieve this they suggest that key global objectives must include reducing the rate of population growth, "promoting lifestyles and consumption of affluent groups and countries that are consistent with ecological sustainability", and "making all individuals and organisations aware of their responsibilities for health and its environmental basis". Their definition of 'ecological sustainability' is interesting here for the way that it emphasises the link between ecological systems and human health: "providing an environment that promotes health – by reducing the risk of physical, chemical and biological hazards and ensuring that everyone has the means to acquire the resources on which health depends".

This conceptual coupling of ecological systems and human physical and mental health is also reflected explicitly in the model of *Green Infrastructure, Ecosystem*

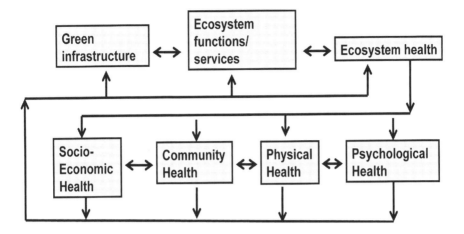

Figure 8.5   Conceptual Framework linking Green Infrastructure, Ecosystem and Human Health

Note: Arrows indicate pathways of influence among different parts of the system.

Source: Adapted from Tzoulas et al. (2007) Figure 1, p. 173.

and Human Health proposed by Tzoulas et al. (2007) (see Figure 8.5). These authors review research that suggests links between the wider ecosystem, including natural environmental and socio-economic processes, and various dimensions of human wellbeing including healthy socio-economic conditions, physical health and psychological health. This kind of whole system perspective helps to contextualise each part of the system in relation to other areas of life. It suggests that it is likely to be difficult to attribute a change in the system solely to variation in one aspect of it. From a public mental health perspective, therefore, this models also suggests that focusing attention primarily on a limited range of specific interventions, where there seems to be evidence that they make a difference in some settings, may not be the most effective strategy to achieve health improvement in the population overall. A wider strategy, that prioritises human health and sustainability in human activities generally, might have more broadly based benefits for wellbeing in the long term.

Prospects for mental health geographies

To conclude, this book has put forward an argument that focuses on the wider determinants of human mental health, which comprise relationships linking individual people to their environment, made up of complex and interrelated material, social and symbolic dimensions. These aspects of environment relate to individual experience in ways which may contribute to good mental health or may increase risks of mental illness. These processes are intricately interconnected, dynamic and difficult to predict with certainty and therefore have several attributes corresponding to aspects of complexity theory. Studies of relationships between people and their environment in different settings can help us to grasp some aspects of this complexity and variety in the wider determinants of mental health, which makes geography, with its emphasis on space and place, a discipline relevant to research in mental public health.

Although a good deal of progress has been made in health geography and related disciplines to understand how space and place matter to mental health, a number of challenges present themselves in terms of the future agenda for research in this field. It seems likely that there will be increased efforts to build temporal as well as spatial perspectives into health geography in order to study dynamic processes, and study how change in environment and change in mental health are related. This will help, for example, to build new knowledge about human experience in varying environments over the life course and its relevance for mental health. New perspectives on space and scale will be required to allow for a more sophisticated understanding of how different parts of complex systems interrelate, and also to allow us to conceptualise imagined and virtual geographies of human experience more completely. This would show us more clearly the meaning of social and material landscapes and their psychological importance for individuals, and would also illuminate the significance of virtual spaces of social interaction and care that have increasing psychological importance in the modern world. Also the importance of issues of sustainability and risk in ecological systems for human psychology will increasingly set future research agendas for geographies of mental health.

Bibliography

Abraham, A., Werning, M., Rakoczy, H., von Cramon, D., and Schubotz, R. (2008). Minds, persons, and space: An fMRI investigation into the relational complexity of higher-order intentionality. *Consciousness and Cognition* 17(2): June: 438–50.

Acierno, R., and Ruggiero, K.J., et al. (2006). Risk and protective factors for psychopathology among older versus younger adults after the 2004 Florida hurricanes. *American Journal of Geriatric Psychiatry*, 14(12): 1051–59.

Adams, R.E., and Boscarino, J.A. (2003). Stress and well-being in the aftermath of the World Trade Center attack: The continuing effects of a communitywide disaster. 53rd Annual Meeting of the Society for the Study of Social Problems. Atlanta, GA: John Wiley & Sons Inc., pp. 175–90.

Adams, R.E., Boscarino, J.A., and Galea, S. (2006). Alcohol use, mental health status and psychological well-being two years after the World Trade Center attacks in New York City. *American Journal of Drug and Alcohol Abuse*, 32(2): 203–24.

Ahern, J., and Galea, S. (2006). Social context and depression after a disaster: the role of income inequality. *Journal of Epidemiology and Community Health*, 60(9): 766–70.

Ahern, M., Kovats, S., Wilkinson, P., Few, R., and Matthies, F. (2005). Global Health Impacts of Floods: Epidemiologic Evidence. *Epidemiologic Reviews*, 27: 36–46.

Ahern, M., and Kovats, S. (2006). The health impacts of floods, in R. Few and F. Matthies (eds) *Flood Hazards and Health*. London, Earthscan, pp. 28–53.

Aitken, S. (2000). Play, rights and borders: gender-bound parents and the social construction of children, in S. Holloway and G. Valentine (eds), *Children's Geographies*. London: Routledge, pp. 119–37.

Albrecht, G., Sartore, G.M., Connor, L., Higginbotham, N., Freeman, S., Kelly, B., et al. (2007). Solastalgia: the distress caused by environmental change. *Australasian Psychiatry*, 15, S95–S98.

Alesina, A., Di Tella, R. and MacCulloch, R. (2004). Inequality and happiness: are Europeans and Americans different? *Journal of Public Economics*, 88(9–10): 2009–42.

Almedom, A.M., and Summerfield, D. (2004). Mental well-being in settings of 'complex emergency': An overview. *Journal of Biosocial Science*, 36(4): 381–8.

Almog, M., Curtis, S., Copeland, A., and Congdon, P. (2004). Geographical variation in acute psychiatric admissions within New York City 1990–2000: growing inequalities in service use? *Social Science and Medicine*, 59(2): 361–76.

Alonso, J., Angermeyer, M., Bernert, R., Bruffaerts, T., Brugha, T., Bryson, H., et al. (2004a). Prevalence of mental disorders in Europe: Results from the European Study of the Epidemiology of Mental Disorders (ESEMeD) project. *Acta Psychiatrica Scandinavica*, 109(Suppl. 420): 21–7.

Alonso, J., Angermeyer, M., Bernert, R., Bruffaerts, T., Brugha, T., Bryson, H., et al. (2004b). Disability and quality of life impact of mental disorders in Europe: Results from the European Study of the Epidemiology of Mental Disorders (ESEMeD) Project. *Acta Psychiatrica Scandinavica*, 109(Suppl. 420): 38–46.

Anderson, K., and Smith, S.J. (2001). Emotional geographies. *Transactions of the Institute of British Geographers*, 26(1): 7–10.

Anderson, K. (2005). Griffith Taylor Lecture, Geographical Society of New South Wales, 2004: Australia and the 'state of nature/native'. *Australian Geographer*, 36(3): 267–82.

Andreasen, N. (2007). DSM and the death of phenomenology in America: an example of unintended consequences. *Schizophrenia Bulletin*, 33(1): 108–12.

Aneshensel, C.S. (2005). Research in mental health: Social etiology versus social consequences. *Journal of Health and Social Behavior*, 46(3): 221–8.

Aneshensel, C.S., and Sucoff, C.A. (1996). The neighborhood context of adolescent mental health. *Journal of Health and Social Behavior*, 37(4): 293–310.

Aniebue, P.N., and Ekwueme, O.C. (2009). Health-seeking behaviour of mentally ill patients in Enugu, Nigeria. *South African Journal of Psychiatry*, 15(1): 19–22.

Appleton, J. (1996). *The Experience of Landscape*. London: Wiley.

Arieti, S. (1962). The microgeny of thought and perception. *Archives of General Psychiatry*, 6, 76–90.

Arshavsky, Y.I. (2006). 'Scientific roots' of dualism in neuroscience. *Progress in Neurobiology*, 79(4): 190–204.

Baer, L. (2001). *Imp of the Mind*. New York: Plume.

Barnett, C., and Land, D. (2007). Geographies of generosity: Beyond the 'moral turn'. *Geoforum*, 38(6): 1065–75.

Barrett, F. (2000). Disease and Geography: the History of an Idea. Toronto: Geographical Monographs, Geography Dept, York University, 47000 Keele St., Toronto, Ontario, Canada M3J IP3.

Barton, H., and Grant, M. (2006). A health map for the local human habitat. *Journal of the Royal Society for Promotion of Health*, 126, 252–3.

Basar, E., and Karakas, S. (2006). Neuroscience is awaiting for a breakthrough: An essay bridging the concepts of Descartes, Einstein, Heisenberg, Hebb and Hayek with the explanatory formulation in this special issue. *International Journal of Psychophysiology*, 60(2): 194–201.

Basar, E. (2006). The theory of the whole-brain-work. *International Journal of Psychophysiology*, 60(2): 133–8.

Bassin, M. (1992). Geographical Determinism in Fin-de-siècle Marxism – Georgii Plekhanov, and the Environmental Basis of Russian History. *Annals of the Association of American Geographers*, 82(1): 3–22.

Bauer, K. (2003). Distributive justice and rural healthcare: a case for e-health. *Int. J. Appl. Philos.*, 17(2): 241–52.

Baumeister, A., Kupstas, F., and Klindworth, L. (1991). The New Morbidity – A National Plan of Action. *American Behavioral Scientist*, 34(4): 468–500.

Beck, U. (1992). *Risk Society: Towards a New Modernity*. London: Sage.

Beck, U. (1999). *World Risk Society*. Cambridge: Polity Press.

Bede (721). *The Life and Miracles of St. Cuthbert, Bishop of Lindesfarne* (translated by J.A. Giles).

Beehler, G., Baker, J., Falkner, K., Chegerova, T., Pryshchepava, A., Chegerov, V., Zevon, M., Bromet, E., Havenaar, J., Valdismarsdottir, H., Moysich, K. B. (2008). A multilevel analysis of long-term psychological distress among Belarusians affected by the Chernobyl disaster. *Public Health*, 122(11): 1239–49.

Bell, C., and Newby, H. (1971). *Community Studies: An Introduction to the Sociology of the Local Community*. London, Allen and Unwin.

Bell, M. (1999). Rehabilitating middle England; integrating ecology, aesthetics and ethics, in A. Williams (ed.), *Therapeutic Landscapes: The Dynamic Between Place and Wellness*. Maryland: University Press of America, pp. 15–28.

Bell, M., and Evans, D.M. (1997). Greening 'the heart of England' – Redemptive science, citizenship, and 'symbol of hope for the nation'. *Environment and Planning D: Society and Space*, 15(3): 257–79.

Bennet, K. (2004). Emotionally intelligent research. *Area*, 36(4): 414–22.

Bennet, L., and Gunn, A.J. (2006). The fetal origins of adult mental illness. *Early Life Origins of Health and Disease*, 573, 204–18.

Bennett, M.R. (2007). Development of the concept of mind. *Australian and New Zealand Journal of Psychiatry*, 41(12): 943–56.

Bentham, J. (1791). Panopticon: or, the inspection-house. Containing the idea of a new principle of construction applicable to any sort of establishment, in which persons...are to be kept...and in particular to penitentiary-houses, prisons, ...In a series of letters, written in...1787...By Jeremy Bentham. Dublin: Thomas Byrne.

Berkman, L.F., Glass, T., Brissette, I., and Seeman, T.E. (2000). From social integration to health: Durkheim in the new millennium. *Social Science and Medicine*, 51(6): 843–57.

Bernard, P., Charafeddine, R., Frohlich, K.L., Daniel, M., Kestens, Y., and Potvin, L. (2007). Health inequalities and place: A theoretical conception of neighbourhood. *Social Science and Medicine*, 65, 1839–52.

Beyers, J., Bates, J.E., Pettit, G.S., Dodge, K.A. (2003). Neighborhood structure, parenting processes, and the development of youths' externalizing behaviors: A multilevel analysis. *American Journal of Community Psychology*, 31(1–2): 35–53.

Bhui, K., Craig, T., Mohamud, S., Warfa, N., Stansfeld, S.A., Thornicroft, G., Curtis, S., McCrone, P. (2006). Mental disorders among Somali refugees. *Social Psychiatry and Psychiatric Epidemiology*, 41(5): 400–408.

Bickerstaff, K., and Walker, G. (2003). The place(s) of matter: matter out of place – public understandings of air pollution. *Progress in Human Geography*, 27(1): 45–67.

Bickerstaff, K. (2004). Risk perception research: socio-cultural perspectives on the public experience of air pollution. *Environment International*, 30(6): 827–40.

Blakely, T.A., and Woodward, A.J. (2000). Ecological effects in multi-level studies. *Journal of Epidemiology and Community Health*, 54(5): 367–74.

Blunt, A., and Dowling, R. (2006). *Home* (Key Ideas in Geography). Oxford: Routledge.

Bondi, L. (1999). Stages on journeys: Some remarks about human geography and psychotherapeutic practice. *Professional Geographer*, 51(1): 11–24.

Bondi, L. (2003). A situated practice for (re)situating selves: trainee counsellors and the promise of counselling. *Environment and Planning A*, 35(5): 853–70.

Bondi, L. (2005). Making connections and thinking through emotions: between geography and psychotherapy. *Transactions of the Institute of British Geographers*, 30(4): 433–48.

Bondi, L. (2009). Counselling in rural Scotland: care, proximity and trust. *Gender Place and Culture*, 16(2): 163–79.

Bondi, L., and Fewell, J. (2003). 'Unlocking the cage door': the spatiality of counselling. *Social and Cultural Geography*, 4(4): 527–47.

Bourdieu, P. (1984). *Distinction: A Social Critique of the Judgement of Taste* (Translated from the 1979 original in French by R. Nice). Cambridge MA: Harvard University Press.

Bourque, L.B., Siegel, J.M., Kano, M., and Wood, M.M. (2006). Weathering the storm: The impact of hurricanes on physical and mental health. *Annals of the American Academy of Political and Social Science*, 604, 129–51.

Bowlby, J. (1969). *Attachment and Loss*. London: Hogarth Press.

Bowling, A. (1997). *Measuring Health: A Review of Quality of life Measurement Scales*. Oxford: Oxford University Press.

Boydell, K.M., Pong, R., Volpe, T., Tilleczek, K., Wilson, E., and Lemieux, S. (2006). Family perspectives on pathways to mental health care for children and youth in rural communities. *Journal of Rural Health*, 22(2): 182–88.

Brace, C., Bailey, A.R., and Harvey, D.C. (2006). Religion, place and space: a framework for investigating historical geographies of religious identities and communities. *Progress in Human Geography*, 30(1): 28–43.

Bracha, H. (2006). Human brain evolution and the "Neuroevolutionary Time-depth Principle:" implications for the reclassification of fear-circuitry-related traits in DSM-V and for studying resilience to warzone-related posttraumatic stress disorder. *Progress in Neuro-Psychopharmacology and Biological Psychiatry*, 30(5): 827–53.

Bracken, P.J., Giller, J.E., and Summerfield, D. (1995). Psychological responses to war and atrocity – the limitations of current concepts. *Social Science and Medicine*, 40(8): 1073–82.

Bray, F. (2005). The Inner Quarters; Oppression or Freedom? in R.G. Knapp and K.Y. Lo (eds), *House, Home, Family: Living and Being Chinese*. Honolulu: University of Hawaii Press.

Bridgen, P. (2006). Social capital, community empowerment and public health: policy developments in the UK since 1997. *Policy and Politics*, 34(1): 27–50.

Brown, A.S., Susser, E.S., Lin, S.P., Neugebauer, R., and Gorman, J.M. (1995). Increased risk of affective-disorders in males after 2nd-trimester prenatal exposure to the Dutch-hunger-winter of 1944–45. *British Journal of Psychiatry*, 166, 601–606.

Brown, A.S., van Os, J., Driessens, C., Hoek, H.W., and Susser, E.S. (2000). Further evidence of relation between prenatal famine and major affective disorder. *American Journal of Psychiatry*, 157(2): 190–5.

Brown, J.D. (2004). Knowledge, uncertainty and physical geography: towards the development of methodologies for questioning belief. *Transactions of the Institute of British Geographers*, 29(3): 367–81.

Brown, S.C., Mason, C.A., Lombard, J.L., Martinez, F., Plater-Zyberk, E., Spokane, A.R., et al. (2009). The relationship of built environment to perceived social support and psychological distress in Hispanic elders: The role of "eyes on the street". *Journals of Gerontology Series B: Psychological Sciences and Social Sciences*, 64(2): 234–46.

Brown, T., McLafferty, S., and Moon, G. (eds) (2009). *Companion to Health and Medical Geography*. Oxford: Blackwell.

Brownlow, A. (2005). A geography of men's fear. *Geoforum*, 36(5): 581–92.

Bullock, T.H. (2006). How do brains evolve complexity? An essay. *International Journal of Psychophysiology*, 60(2): 106–109.

Burke, J., O'Campo, P., Salmon, C., and Walker, R. (2009). Pathways connecting neighborhood influences and mental well-being: Socioeconomic position and gender differences. *Social Science and Medicine*, 68(7): 1294–1304.

Butler, R., and Parr, H. (eds) (1999). *Mind and Body Spaces*. London: Routledge.

Buttimer, A. (2006). Afterword: Reflections on geography, religion, and belief systems. *Annals of the Association of American Geographers*, 96(1): 197–202.

Byrne, D. (1998). *Complexity Theory and the Social Sciences*. London: Routledge.

Callard, F. (2006). 'The sensation of infinite vastness'; or, the emergence of agoraphobia in the late 19th century. *Environment and Planning D: Society and Space*, 24(6): 873–89.

Campbell, C., Cornish, F., and McLean, C. (2004). Social capital, participation and the perpetuation of health inequalities: Obstacles to African–Caribbean participation in 'partnerships' to improve mental health. *Ethnicity and Health*, 9(4): 313–35.

Campbell, J. (1994). *Past, Space and Self*. Cambridge MA: MIT Press.

Campbell, P., Wright, J., Oyebode, J., Job, D., Crome, P., Bentham, P., et al. (2008). Determinants of burden in those who care for someone with dementia. *International Journal of Geriatric Psychiatry*, 23(10): 1078–85.

Care Services Improvement Partnership North West. (undated). Mental Health Impact Assessment.

Carlisle, S., and Hanlon, P. (2007). Well-being and consumer culture: a different kind of public health problem? *Health Promotion International*, pp. 261–68.

Carpenter, M. (2000). 'It's a small world': mental health policy under welfare capitalism since 1945. *Sociology of Health and Illness*, 22(5): 602–20.

Carpiano, R. M. (2006). Toward a neighborhood resource-based theory of social capital for health: Can Bourdieu and sociology help? *Social Science and Medicine*, 62(1): 165–75.

Carpiano, R.M. (2007). Neighborhood social capital and adult health: An empirical test of a Bourdieu-based model. *Health and Place*, 13(3): 639–55.

Carpiano, R.M. (2008). Actual or potential neighborhood resources and access to them: Testing hypotheses of social capital for the health of female caregivers. *Social Science and Medicine*, 67(4): 568–82.

Carpiano, R.M. (2009). Come take a walk with me: The ;Go-Along; interview as a novel method for studying the implications of place for health and well-being. *Health and Place*, 15(1): 263–72.

Carroll, B., Morbey, H., Balogh, R., and Araoz, G. (2009). Flooded homes, broken bonds, the meaning of home, psychological processes and their impact on psychological health in a disaster. *Health and Place*, 15(2): 540–7.

Carstairs, V. (1995). Deprivation indices: Their interpretation and use in relation to health. *Journal of Epidemiology and Community Health*, 49, S3–S8.

Carstairs, V.R.M. (1990). Deprivation and Health in Scotland UK. *Health Bulletin (Edinburgh)*, 48(4): 162–75.

Cartier, C. (2003). China's living houses: Folk beliefs, symbols, and household ornamentation. *Annals of the Association of American Geographers*, 93(2): 527–8.

Castree, N. (2004). Differential Geographies: place, indigenous rights and 'local' resources. *Political Geography*, 23, 133–67.

Cattell, V. (2001). Poor people, poor places, and poor health: the mediating role of social networks and social capital. *Social Science and Medicine*, 52(10): 1501–16.

Cattell, V. (2004a). Having a laugh and mucking in together: Using social capital to explore dynamics between structure and agency in the context of declining and regenerated neighbourhoods. *Sociology: The Journal of the British Sociological Association*, 38(5): 945–63.

Cattell, V. (2004b). Social networks as mediators between the harsh circumstances of people' lives and their lived experience of health and well-being, in C. Phillipson, G. Allen, and D. Morgan (eds), *Social Networks and Social Exclusion*, Aldershot: Ashgate, pp. 142–61.

Cattell, V., Dines, N., Gesler, W., and Curtis, S. (2008). Mingling, observing, and lingering: Everyday public spaces and their implications for well-being and social relations. *Health and Place*, 14(3): 544–61.

Chalmers, D. (ed.) (2002). *Philosophy of Mind: Classical and Contemporary Readings*. Oxford: Oxford University Press.

Chapman, J. (1966). The early symptoms of schizophrenia. *British Journal of Medical Psychology* (112): 225–51.

Chatwin, B. (1988). *The Songlines*. London: Penguin.

Chen, A., Keith, V., Leong, K., Airriess, C., Li, Chung, W., and Lee, C. (2007). Hurricane Katrina: prior trauma, poverty and health among Vietnamese–American survivors *International Nursing Review*, 54(4): 324–31.

Cherniack, E.P., Troen, B.R., Florez, H.J., Roos, B.A., and Levis, S. (2009). Some new food for thought: the role of vitamin D in the mental health of older adults. *Curr. Psychiatry Rep.*, 11(1): 12–19.

Chiou, S., and Krishnamurti, R. (1997). Unraveling feng-shui. *Environment and Planning B: Planning and Design*, 24(4): 549–72.

Christensen, P., James, A., and Jenks, C. (2000). Home and movement: children constructing 'family time', in S. Holloway, and G. Valentine (eds), *Children's Geographies*. London: Routledge, pp. 139–55.

Chu, T.Q., Seery, M.D., Ence, W.A., Holman, E.A., and Silver, R.C. (2006). Ethnicity and gender in the face of a terrorist attack: A national longitudinal study of immediate responses and outcomes two years after September 11. *Basic and Applied Social Psychology*, 28(4): 291–301.

Cifuentes, M., Sembaljwe, G., Tak, S., Gore, R., Kriebel, D., and Punnett, L. (2008). The association of major depressive episodes with income inequality and the human development index. *Social Science and Medicine*, 67(4): 529–39.

Coggins, T., Cooke, A., Friedli, L, Nicholls, J., Scott-Samuel, A., Stansfield, J. (2007). Mental Well-Being Impact Assessment: A Toolkit Care Services Improvement Partnership North West.

Cohen, D.A., Farley, T.A., and Mason, K. (2003). Why is poverty unhealthy? – Social and physical mediators. *Social Science and Medicine*, 57(9): 1631–41.

Cohen, D.A., Inagami, S., and Finch, B. (2008). The built environment and collective efficacy. *Health and Place*, 14(2): 198–208.

Cohen, M. (2007). Environmental toxins and health – The health impact of pesticides. *Australian Family Physician*, 36(12): 1002–04.

Coldefy, M., and Curtis, S. (2009). French Asylum Geography: A Spatial Innovation Diffusion Analysis. Presented at the International Medical Geography Symposium, McMaster University, Hamilton, Canada.

Coleman, J. (1988). Social capital and the creation of human capital. *American Journal of Sociology*, 94, S95–S120.

Collins, J., Winefield, H., Ward, L., Turnbull, D. (2008). Barriers to seeking help for mental health problems in rural Australia. *International Journal of Psychology*, 43(3–4): 668–9.

Collins, J.E., Winefield, H., Ward, L., and Turnbull, D. (2009). Understanding help seeking for mental health in rural South Australia: thematic analytical study. *Australian Journal of Primary Health*, 15(2): 159–65.

Collishaw, S., B. Maughan, et al. (2004). Time trends in adolescent mental health. *Journal of Child Psychology and Psychiatry*, 45(8): 1350–62.

Commission on Mental Health. (2003). President's New Freedom Commission on Mental Health Rockville, Maryland.

Congdon, P. (1996). Suicide and parasuicide in London: a small area study. *Urban Studies*, 33(1): 137–58.

Congdon, P. (1997). Multilevel and clustering analysis of health outcomes in small areas. *European Journal of Population [Revue Europeenne De Demographie]*, 13(4): 305–38.

Congdon, P. (2006). Estimating population prevalence of psychiatric conditions by small area with applications to analysing outcome and referral variations. *Health and Place*, 12, 465–78.

Conradson, D. (2003). Spaces of care in the city: the place of a community drop-in centre *Social and Cultural Geography*, 4(4): 507–25.

Conradson, D. (2005). Landscape, care and the relational self: therapeutic encounters in rural England. *Health and Place*, 11(4): 337–48.

Coombes, P., and Barber, K. (2005). Environmental determinism in Holocene research: causality or coincidence? *Area*, 37(3): 303–11.

Cresswell, T. (1997). Weeds, plaques and bodily secretions; a geography of interpretation of metaphors and displacement. *Annals of Association of American Geographers*, 87(2): 330–45.

Cresswell, T. (1999). Embodiment, power and the politics of mobility: the case of female tramps and hobos. *Transactions of Institute of British Geographers*, NS 24, 175–92.

Crighton, E.J., Elliott, S.J., van der Meer, J., Small, I., and Upshur, R. (2003). Impacts of an environmental disaster on psychosocial health and well-being in Karakalpakstan. *Social Science and Medicine*, 56(3): PII S0277–9536(0202)00054–00050.

Cryder, C., Kilmer, R., Tedeschi, R., and Calhoun, L. (2006). An explanatory study of post-traumatic growth in children following a natural disaster. *American Journal of Orthopsychiatry*, 76(1): 65–9.

Cummins, S., Curtis, S., Diez-Roux, A., and Macintyre, S. (2007). Understanding and representing 'place' in health research: A relational approach. *Social Science and Medicine*, 65, 1825–38.

Cummins, S., Macintyre, S., Davidson, S., and Ellaway, A. (2005). Measuring neighbourhood social and material context: generation and interpretation of ecological data from routine and non-routine sources. *Health and Place*, 11(3): 249–60.

Curtis, S., and Taket, A. (1996). *Health and Societies: Changing Perspectives*. London: Arnold.

Curtis, S., and Rees-Jones, I. (1998). Is there a place for geography in the analysis of health inequality? *Sociology of Health and Illness*, 20(5): 645–72.

Curtis, S. (2004). *Health and Inequality: Geographical Perspectives*. London: Sage.

Curtis, S., and Cummins, S. (2007). Ecological studies, in S.E. Galeo (ed.), *Macro Social Determinants of Health*. New York: Springer, pp. 327–48.

Curtis, S. (2007). Socio-economic status and geographies of psychiatric inpatient service use. Places, provision, power and wellbeing. *Epidemiologia E Psichiatria Sociale – An International Journal for Epidemiology and Psychiatric Sciences*, 16(1): 10–15.

Curtis, S. (2008). How can we address health inequality through healthy public policy in Europe? *European Urban and Regional Studies*, 15(4): 293–305.

Curtis, S., Daya, S., Khatib, Y, Pain, R., Rothon, C., Stansfeld, S. (2009). Mapping links between young people, neighbourhoods, schools and families, with respect to adolescent mental health (Report to the Nuffield Foundation). Durham: Durham University, Geography Dept.

Curtis, S., and Nadiruzzama (2009). The mental health impacts of major floods: a perspective from medical geography. Durham: Institute of Hazard and Risk Research, Durham University.

Curtis, S., Gesler, W., Priebe, S., Francis, S. (2009). New spaces of inpatient care for people with mental illness: A complex 'rebirth' of the clinic? *Health and Place*, 15(1): 340–8.

Curtis, S., Setia, M., and Quesnel-Vallée, A. (2010). Socio-Geographic Mobility and Health Status: A longitudinal analysis using the National Population Health Survey of Canada. *Social Science & Medicine*, 69(12): 1845–1853.

Curtis, S., and Riva, M. (2010a). Complexity in human health and healthcare: recent advances in geographies of human health. *Progress in Human Geography*, 34(2): 215–223.

Curtis, S. and Riva, M. (2010b). Complexity and geographies of health care systems and policy. *Progress in Human Geography* (forthcoming).

Curtis, S., Congdon, P., Almog, M., and Ellermann, R. (2009). County variation in use of inpatient and ambulatory psychiatric care in New York State 1999–2001: Need and supply influences in a structural model. *Health and Place*, 15(2): 568–77.

Curtis, S., Copeland, A., Fagg, J., Congdon, P., Almog, M., and Fitzpatrick, J. (2006). The ecological relationship between deprivation, social isolation and rates of hospital admission for acute psychiatric care: a comparison of London and New York City. *Health and Place*, 12(1): 19–37.

Curtis, S., Gesler, W., Fabian, K., Francis, S., and Priebe, S. (2007). Therapeutic landscapes in hospital design: a qualitative assessment by staff and service users of the design of a new mental health inpatient unit. *Environment and Planning C: Government and Policy*, 25(4): 591–610.

Cutchin, M.P. (2002). Virtual medical geographies: conceptualizing telemedicine and regionalization. *Progress in Human Geography*, 26(1): 19–39.

Dallaire, B., McCubbin, M., Morin, P., Cohen, D. (2000). Civil commitment due to mental illness and dangerousness: the union of law and psychiatry within a treatment-control system. *Sociology of Health and Illness*, 22(5): 679–99.

Dambrun, M., Kamiejski, R., Haddadi, N., and Duarte, S. (2009). Why does social dominance orientation decrease with university exposure to the social sciences? The impact of institutional socialization and the mediating role of "geneticism". *European Journal of Social Psychology*, 39(1): 88–100.

Dauncey, K., Giggs, J., Baker, K., and Harrison, G. (1993). Schizophrenia in Nottingham – Lifelong Residential Mobility of a Cohort. *British Journal of Psychiatry*, 163, 613–19.

Davidson, J. (2000). A phenomenology of fear: Merleau-Ponty and agoraphobic life-worlds – Paper is dedicated to Jim Davidson, 1965–2000. *Sociology of Health and Illness*, 22(5): 640–60.

Davidson, J. (2003). *Phobic Geographies: The Phenomenology and Spatiality of Identity*. Aldershot: Ashgate.

Davidson, J., and Milligan, C. (2004). Embodying emotion, sensing space: introducing emotional geographies,. *Social and Cultural Geography*, 5(4): 523–32.

Davidson, J., Bondi, L., and Smith, M. (2005). *Emotional Geographies*. Aldershot: Ashgate.

Davidson, P.W., Myers, G.J., and Schroeder, S.R. (1997). Overview: Environmental contaminants and developmental disabilities. *Mental Retardation and Developmental Disabilities Research Reviews*, 3(3): 221–2.

Davies, B. (1968). *Social Needs and Resources*. London: Joseph.

Day, R. (2008). Local environments and older people's health: Dimensions from a comparative qualitative study in Scotland. *Health and Place*, 14(2): 299–312.

De Botton, A. (2004). *Status Anxiety*. London: Penguin.

De Botton, A. (2006). *The Architecture of Happiness*. London: Penguin Books.

De Groot, W., and Van den Born, R. (2003). Visions of Nature and landscape type preferences: an exploration in the Netherlands. *Landscape and Urban Planning*, 63, 127–38.

De Silva, M.J., McKenzie, K., Harpham, T., and Huttly, S.R.A. (2005). Social capital and mental illness: a systematic review. *Journal of Epidemiology and Community Health*, 59(8): 619–27.

De Silva, P., and Rachman, S. (1998). *Obsessive-Compulsive Disorder: The Facts*. Oxford: Oxford University Press.

De Vries, S., Verheij, R.A., Groenewegen, P.P., and Spreeuwenberg, P. (2003). Natural environments – healthy environments? An exploratory analysis of the relationship between greenspace and health. *Environment and Planning A*, 35(10): 1717–31.

Dean, K., and James, H. (1981). Social factors and admission to psychiatric hospital: schizophrenia in Plymouth. *Transactions of Institute of British Geographers*, NS 6, 39–52.

Dear, M., and Taylor, S. (1982). *Not On Our Street: Community Attitudes To Mental Health Care*. London: Pion.

Dear, M., and Wolch, J. (1987). *Landscapes of Despair: From Institutionalisation to Homelessness*. Oxford: Polity.

Dekker, J., Peen, J., Goris, A., Heijnen, H., and Kwakman, H. (1997). Social deprivation and psychiatric admission rates in Amsterdam. *Social Psychiatry and Psychiatric Epidemiology*, 32(8): 485–92.

Dematteis, G. (2008). New approaches of human geography in a nonlinear history. *Quaderni Storici*, 43(1): 15.

Department of Health. (2005). *Resource Allocation Weighted Capitation Formula. Fifth Edition*. London: Department of Health.

Department of Health. (2008). *Resource Allocation Weighted Capitation Formula. Sixth Edition*. London: Department of Health.

DEPFRA (2008). *Sustainable Development Indicators in Your Pocket 2008*. London: Department of Food and Rural Affairs.

Descartes, R. (1985). *Meditations on First Philosophy*. Translated by J. Cottingham, First published posthumously by Camusat and Petit, Paris 1657. Cambridge: Cambridge University Press.

Descartes, R. (2002). Meditations on first Philosophy. Second Meditation: The nature of the Human Mind, and How It is Better Known than the Body, and Sixth Meditation: The Existence of Material Things, and the Real Distinction between Mind and Body. Reproduced from Descartes (1985). In D. Chalmers (ed.), *Philosophy of Mind: Classical and Contemporary Readings*, pp. 10–21.

DeVerteuil, G., Hinds, A., Lix, L., Walker, J., Robinson, R., and Roos, L.L. (2007). Mental health and the city: Intra-urban mobility among individuals with schizophrenia. *Health and Place*, 13(2): 310–23.

Di Mattei, V.E., Prunas, A., Novella, L., Marcone, A., Cappa, S., and Sarno, L. (2008). The burden of distress in caregivers of elderly demented patients and its relationship with coping strategies. *Neurological Sciences*, 29(6): 383–389.

Dongchu, H. (1991). *The Way of the Virtuous: The Influence of Art and Philosophy on Chinese Garden Design*. Bejing, China: New World Press.

Douglas, M.J., Conway, L., Gorman, D., Gavin, S., and Hanlon, P. (2001). Developing principles for health impact assessment. *Journal of Public Health Medicine*, 23(2): 148–54.

Drukker, M., Kaplan, C., Feron, F., and van Os, J. (2003). Children's health-related quality of life, neighbourhood socio-economic deprivation and social capital. A contextual analysis. *Social Science and Medicine*, 57(5): 825–41.

Drukker, M., and van Os, J. (2003). Mediators of neighbourhood socioeconomic deprivation and quality of life. *Social Psychiatry and Psychiatric Epidemiology*, 38(12): 698–706.

Dubow, E., Edwards, S., Ippolito, M.F. (1997). Life stressors, neighborhood disadvantage, and resources: A focus on inner-city children's adjustment. *Journal of Clinical Child Psychology*, 26(2): 130–44.

Duncan, C., Jones, K., Moon, G. (1995). Psychiatric Morbidity – A Multilevel Approach To Regional Variations in The UK. *Journal of Epidemiology and Community Health*, 49(3): 290–5.

Durkheim, E. (1993). *Suicide: A Study in Sociology* (Translated by J. Spaulding, and G. Simpson) from the original by Durkheim (1858–1917). London: Routledge.

Dutton, B. (2006). Benedict De Spinoza (1632–1677), in J. Fieser and B. Dowden (eds) *Internet Encyclopedia of Philosophy*, http://www.iep.utm.edu/spinoza/#/SSH3a.ii. Accessed on 30.05.2010.

Dyck, I. (2006). Travelling tales and migratory meanings: South Asian migrant women talk of place, health and healing. *Social and Cultural Geography*, 7(1): 1–18.

Dyck, I., and Dossa, P. (2007). Place, health and home: Gender and migration in the constitution of healthy space. *Health and Place*, 13(3): 691–701.

Dyck, I., Kontos, P., Angus, J., and McKeever, P. (2005). The home as a site for long-term care: meanings and management of bodies and spaces. *Health and Place*, 11(2): 173–85.

Edginton, B. (1997). Moral architecture: the influence of the York Retreat on asylum design. *Health and Place*, 3(2): 91–100.

Egede, L.E., Frueh, C.B., Richardson, L.K., Acierno, R., Mauldin, P.D., Knapp, R.G., et al. (2009). Rationale and design: telepsychology service delivery for depressed elderly veterans. *Trials*, 10, 14.

Eibner, C., Sturn, R., and Gresenz, C.R. (2004). Does relative deprivation predict the need for mental health services? *J Ment Health Policy Econ*, 7(4): 167–75.

Ellaway, A., McKay, L., Macintyre, S., Kearns, A., and Hiscock, R. (2004). Are social comparisons of homes and cars related to psychosocial health? *International Journal of Epidemiology*, 33(5): 1065–71.

Ernste, H. (2004). The pragmatism of life in poststructuralist times. *Environment and Planning A*, 36(3): 437–50.

Ertel, K.A., Koenen, K.C., and Berkman, L.F. (2008). Incorporating home demands into models of job strain: Findings From the Work, Family, and Health Network. *Journal of Occupational and Environmental Medicine*, 50(11): 1244–52.

Estby, S.N., Freel, M.I., Hart, L.K., Reese, J.L., and Clow, T.J. (1994). A Delphi study of the basic principles and corresponding care goals of holistic nursing practice. *J Holist Nurs*, 12(4): 402–13.

Evans, G., Wells, N., Chan, E., Saltzman, H. (2000). Housing and mental health *Journal of Consulting and Clinical Psychology*, 68(3): 526–30.

Evans, G., Wells, N., Moch, A. (2003). Housing and mental health: A review of the evidence and a methodological and conceptual critique. *Journal of Social Issues*, 59(3): 475–500.

Evans, G., and Lapore, S. (2008). Psychosocial processes linking the environment and mental health, in H. Freeman and S. Stansfeld (eds), *The Impact of the Environment on Psychiatric Disorder*. Hove, Sussex, UK: Routledge.

Evans, J., Hyndman, S., Stewart-Brown, S., Smith, D., and Petersen, S. (2000). An epidemiological study of the relative importance of damp housing in relation to adult health. *Journal of Epidemiology and Community Health*, 54(9): 677–86.

Evans, J., Middleton, N., and Gunnell, D. (2004). Social fragmentation, severe mental illness and suicide. *Social Psychiatry and Psychiatric Epidemiology*, 39(3): 165–70.

National Institute for Health and Clinical Excellence (NICE) (2007). Public health interventions to promote positive mental health and prevent mental health disorders among adults: evidence briefing. London: NICE.

Fagg, J., Curtis, S., Clark, C., Congdon, P., and Stansfeld, S.A. (2008). Neighbourhood perceptions among inner-city adolescents: Relationships with their individual characteristics and with independently assessed neighbourhood conditions. *Journal of Environmental Psychology*, 28(2): 128–42.

Fagg, J., Curtis, S., Stansfeld, S., and Congdon, P. (2006). Psychological distress among adolescents, and its relationship to individual, family and area characteristics in East London. *Social Science and Medicine*, 63(3): 636–48.

Fagg, J., Curtis, S., Stansfeld, S.A., Cattell, V., Tupuola, A.M., and Arephin, M. (2008). Area social fragmentation, social support for individuals and psychosocial health in young adults: Evidence from a national survey in England. *Social Science and Medicine*, 66(2): 242–54.

Faris R. and Dunham, H. (1939). *Mental Disorders in Urban Areas*. Chicago: University of Chicago Press.

Farrell, S.P., Mahone, I.H., Zerull, L. M., Guerlain, S., Akan, D., Hauenstein, E., et al. (2009). Electronic screening for mental health in rural primary care: implementation. *Issues Ment Health Nurs*, 30(3): 165–73.

Fauth, R.C., Leventhal, T., and Brooks-Gunn, J. (2005). Early impacts of moving from poor to middle-class neighborhoods on low-income youth. *Journal of Applied Developmental Psychology*, 26(4): 415–39.

Fauth, R.C., Leventhal, T., and Brooks-Gunn, J. (2007). Welcome to the neighborhood? Long-term impacts of moving to low-poverty neighborhoods on poor children's and adolescents' outcomes. *Journal of Research on Adolescence*, 17(2): 249–83.

Fauth, R.C., Leventhal, T., and Brooks-Gunn, J. (2008). Seven years later: Effects of a neighborhood mobility program on poor black and Latino adults' well-being. *Journal of Health and Social Behavior*, 49(2): 119–30.

Fawole, O.I. (2008). Economic violence to women and girls – Is it receiving the necessary attention? *Trauma Violence and Abuse*, 9(3): 167–77.

Ferguson, K.M. (2006). Social capital and children's wellbeing: a critical synthesis of the international social capital literature. *International Journal of Social Welfare*, 15(1): 2–18.

Few R. and Matthias, F. (ed.) (2006). *Flood Hazards and Health*. London: Earthscan.

Firebaugh, G. (2009). Commentary: Is the Social World Flat? W.S. Robinson and the Ecologic Fallacy. *International Journal of Epidemiology*, 38(2): 368–70.

Fleuret, S., and Thouez, J-P. (eds) (2007). *Géographie de la Santé: un Panorama.* Paris: Economica.

Fleuret, S., and Atkinson, S. (2007). Wellbeing, health and geography: A critical review and research agenda. *New Zealand Geographer*, 63(2): 106–18.

Floate, H.F.G. (1969). Mauritian Malaria Epidemic 1866–1868 Geographical Determinism 100 Years Ago. *Journal of Tropical Geography*, 29 (December), 10–20.

Flower, A., Lewith, G.T., and Little, P. (2007). Seeking an oracle: using the Delphi process to develop practice guidelines for the treatment of endometriosis with Chinese herbal medicine. *Journal of Alternative and Complementary Medicine*, 13(9): 969–76.

Foley, R., and Platzer, H. (2007). Place and provision: Mapping mental health advocacy services in London. *Social Science and Medicine*, 64(3): 617–32.

Folkman, S., Lazarus, R., Dunkel-Schetter, C., DeLongis, A., Gruen, R. (1986). Dynamics of a stressful ecounter: cognitive appraisal, coping, and encounter outcomes. *Journal of Personality and Social Psychology*, 50(2): 992–1003.

Fone D., D.F., Lloyd K., Williams G., Watkins J., Palmer S. (2007). Does social cohesion modify the association between area income deprivation and mental health? A multilevel analysis. *International Journal of Epidemiology*, 36(2): 338–45

Ford, T., Goodman, R., and Meltzer, H. (2004). The relative importance of child, family, school and neighbourhood correlates of childhood psychiatric disorder. *Social Psychiatry and Psychiatric Epidemiology*, 39(6): 487–96.

Foucault, M. (1967). *Madness and Civilization: A History of Insanity in the Age of Reason* (First published 1961 as *Histoire de la folie*). London: Routledge.

Foucault, M. (1973). *The Birth of the Clinic* (First Published 1963 as *La naissance de la clinique*), London: Routledge.

Foucault, M. (1977). Discipline and Punish: The Birth of the Prison (First Published in 1975 as *Surveiller et Punir, Naissance de la Prison*). London: Penguin Books.

Foucault, M. (1989). *The Birth of the Clinic: An Archaeology of Medical Perception.* London: Routledge.

Fowler, J.H., and Christakis, N.A. (2009). Dynamic spread of happiness in a large social network: longitudinal analysis over 20 years in the Framingham Heart Study. *British Medical Journal*, 338, 23–27.

Francis, S., Glanville, R., Noble, A., and Scher, P. (1999). *50 Years of Ideas in Health Care Buildings.* London: Nuffield Trust.

Francis, S., and Glanville, R. (2001). *Building a 2020 Vision: Future Health Care Environments.* London: Nuffield Trust.

Franzini, L., Caughy, M.O., Nettles, S.M., and O'Campo, P. (2008). Perceptions of disorder: Contributions of neighborhood characteristics to subjective perceptions of disorder. *Journal of Environmental Psychology*, 28(1): 83–93.

Freedy, J.R., Kilpatrick, D.G., and Resnick, H.S. (1993). Natural Disasters and Mental-health – Theory, Assessment, and Intervention. *Journal of Social Behavior and Personality*, 8(5): 49–103.

Freedy, J.R., Shaw, D.L., Jarrell, M.P., and Masters, C.R. (1992). Towards an Understanding of the Psychological Impact of Natural Disasters – An Application of the Conservation Resources Stress Model. *Journal of Traumatic Stress*, 5(3): 441–54.

Freeman, H. (2008). Housing and mental health, in H. Freeman and S. Stansfeld (eds), *The Impact of the Environment on Psychiatric Disorder*. Hove, Sussex, UK: Routledge.

French, D. (2009). Residential segregation and health in Northern Ireland. *Health and Place*, 15(3): 888–96.

Frimel, T.J. (2001). Florence Nightingale, Martha Rogers and the art of Feng Shui. *Beginnings*, 21(1): 10.

Frohlich, K.L. (2007). Psycho-social correlates of health and health behaviours: What does the term psychosocial mean for public health? *International Journal of Public Health*, 52(1): 2–3.

Fromm, E. (1973). *The Anatomy of Human Destructiveness*. New York, Holt.

Freud, S. (1991). *On Metapsychology – The Theory of Psychoanalysis: "Beyond the Pleasure Principle", "Ego and the Id" and Other Works*. Harmondworth: Penguin Books Ltd.

Freud, S. (1997). *The Interpretation of Dreams*. Wordsworth Editions Limited, New edition.

Frumkin, H. (2003). Healthy places: Exploring the evidence. *American Journal of Public Health*, 93(9): 1451–56.

Fujiwara, T., and Kawachi, I. (2008). Social capital and health – A study of adult twins in the US. *American Journal of Preventive Medicine*, 35(2): 139–44.

Fulkerson, G.M., and Thompson, G.H. (2008). The Evolution of a Contested Concept: A Meta-Analysis of Social Capital Definitions and Trends (1988–2006). *Sociological Inquiry*, 78(4): 536–57.

Fullilove, M. (1996). Psychiatric implications of displacement: contributions from the psychology of place. *American Journal of Psychiatry*, 153(12): 1516–23.

Fullilove, M. (2005). *Root Shock: How Tearing up City Neighborhoods Hurts America, and What We Can Do About It*. New York: Random House.

Galea, S., Maxwell, A.R., and Norris, F. (2008). Sampling and design challenges in studying the mental health consequences of disasters. *International Journal of Methods in Psychiatric Research*, 17, S21–S28.

Galea, S., Nandi, A., and Vlahov, D. (2005). The epidemiology of post-traumatic stress disorder after disasters. *Epidemiologic Reviews*, 27, 78–91.

Galea, S., Vlahov, D., Resnick, H., Ahern, J., Susser, E., Gold, J., et al. (2003). Trends of probable post-traumatic stress disorder in New York City after the September 11 terrorist attacks. *American Journal of Epidemiology*, 158(6): 514–24.

Gatrell, A. (2002). *Geographies of Health*. Oxford: Blackwell.

Gatrell, A., Popay, J., and Thomas, C. (2004). Mapping the determinants of health inequalities in social space: can Bourdieu help us? *Health and Place*, 10(3): 245–57.

Gely, S. (2005). Places, bodies, costumes. Environmental determinism and ancient ethnography. *Latomus*, 64(3): 776–78.

Georges, J., Jansen, S., Jackson, J., Meyrieux, A., Sadowska, A., and Selmes, M. (2008). Alzheimer's disease in real life – the dementia carer's survey. *International Journal of Geriatric Psychiatry*, 23(5): 546–51.

Gesler, W. (1996). Lourdes: healing in a place of pilgrimage. *Health and Place*, 2(2): 95–105.

Gesler, W. (1998a). Bath's reputation as a healing place, in R. Kearns and W. Gesler (eds), *Putting Health into Place; Landscape, Identity and Well-being*. New York: Syracuse University Press.

Gesler, W. (1998b). Sacred spaces, sacred places: The geography of pilgrimages. *Annals of the Association of American Geographers*, 88(3): 535–37.

Gesler, W. (2000). Hans Castorp's journey-to-knowledge of disease and health in Thomas Mann's *The Magic Mountain. Health and Place*, 6(2): 125–34.

Gesler, W. (2003). *Healing Places*. Lanham, MD: Rowman and Littlefield.

Gesler, W. (2005). Therapeutic landscapes: An evolving theme. *Health and Place*, 11(4): 295–97.

Gesler, W., and Curtis, S. (2007). Application of concepts of therapeutic landscapes to the design of hospitals in the UK: the example of a mental health facility in East London, in A. Williams (ed.), *Therapeutic Landscapes*. Aldershot: Ashgate, pp. 149–64.

Gesler, W., Arcury, T.A., and Koenig, H.G. (2000). An introduction to three studies of rural elderly people: effects of religion and culture on health. *J Cross Cult Gerontol*, 15(1): 1–12.

Gesler, W., Bell, M., Curtis, S., Hubbard, P., and Francis, S. (2004). Therapy by design: evaluating the UK hospital building program. *Health and Place*, 10(2): 117–28.

Gesler, W.M. (1992). Therapeutic Landscapes – Medical Issues in Light of the New Cultural-Geography. *Social Science and Medicine*, 34(7): 735–46.

Gesler, W.M. (1993). Therapeutic Landscapes – Theory and a Case-study of Epidauros, Greece. *Environment and Planning D: Society and Space*, 11(2): 171–89.

Gesler, W.M. (1999). Words in wards: language, health and place. *Health and Place*, 5(1): 13–25.

Gibson, J. (1977). The theory of affordances. in R. Shaw, and Bransford, J. (eds), *Perceiving, Acting and Knowing: Toward an Ecological Psychology*. Hillsdale, NJ: Erlbaum, pp. 67–82.

Gibson, J. (1986). *The Ecological Approach to Visual Perception*. Hillsdale, NJ: Erlbaum.

Giddens, A. (1994). *Reflexive Modernization: Politics, Tradition and Aesthetics in the Modern Social Order*. Cambridge: Polity.

Giggs, J. (1973). The distribution of schizophrenics in Nottingham. *Transactions of the Institute of British Geographers* 59, 55–76.

Giggs, J. (1975). The distribution of schizophrenics in Nottingham: a reply. *Transactions of the Institute of British Geographers*, 64, 150–6.

Giggs, J.A. (1986). Mental-disorders and Ecological Structure in Nottingham. *Social Science and Medicine*, 23(10): 945–61.

Giggs, J.A., and Cooper, J.E. (1987). Ecological Structure and the Distribution of Schizophrenia and Affective Psychoses in Nottingham. *British Journal of Psychiatry*, 151, 627–33.

Ginexi, E.M., Weihs, K., Simmens, S.J., and Hoyt, D.R. (2000). Natural disaster and depression: A prospective investigation of reactions to the 1993 Midwest Floods. *American Journal of Community Psychology*, 28(4): 495–518.

Glass, T.A., DeLeon, C.F.M., Seeman, T.E., and Berkman, L.F. (1997). Beyond single indicators of social networks: A LISREL analysis of social ties among the elderly. *Social Science and Medicine*, 44(10): 1503–17.

Gleeson, B. (2008). Critical Commentary. Waking from the Dream: An Australian Perspective on Urban Resilience. *Urban Studies*, 45(13): 2653–68.

Glover, G. (1997). The Mental Illness Needs Index. *Epidemiol Psichiatr Soc.* 6(1 Suppl.), 13–20.

Glover, G., Arts, G., and Wooff, D. (2004). A needs index for mental health care in England based on updatable data. *Social Psychiatry and Psychiatric Epidemiology*, 39(9): 730–8.

Glover, G.R., Robin, E., Emami, J., and Arabscheibani, G.R. (1998). A needs index for mental health care. *Social Psychiatry and Psychiatric Epidemiology*, 33(2): 89–96.

Goeres, M. (1998). Surviving on metaphor: how 'health = hot springs'' created and sustained a town, in R. Kearns and W. Gesler (eds), *Putting Health into Place; Landscape, Identity and Well-being.* New York: Syracuse University Press.

Goldmanrakic, P.S. (1994). Working-memory Dysfunction in Schizophrenia. *Journal of Neuropsychiatry and Clinical Neurosciences*, 6(4): 348–57.

Goldman-Rakic, P.S. (1992). Working Memory And The Mind. *Scientific American*, 267(3): 110–17.

Goodwin, V., and Happell, B. (2007). Consumer and carer participation in mental health care: the carer's perspective: part 2 – barriers to effective and genuine participation. *Issues Ment Health Nurs*, 28(6): 625–38.

Gordon, E.E. (1996). The placebo: An insight into mind-body interaction. *Headache Quarterly – Current Treatment and Research*, 7(2): 117–25.

Goto, T., Wilson, J.P., Kahana, B., and Slane, S. (2006). The Miyake Island volcano disaster in Japan: Loss, uncertainty, and relocation as predictors of PTSD and depression. *Journal of Applied Social Psychology*, 36(8): 2001–26.

Graham, S., and Healey, P. (1999). Relational concepts of space and place: issues for planning theory and practice. *European Planning Studies*, 7, 623–46.

Green, B., Korol, M., Grace, G., Vary, M., Leonard., A. Gleser, G., and Smitons-Cohen, S. (1991). Children and disaster: Age, gender and parental effects on

PTSD symptoms. *Journal of American Academy of Child and Adolescent Psychiatry*, 30(6): 945–51.

Greenho, J.G. (1994). Gibsons Affordances. *Psychological Review*, 101(2): 336–42.

Gregory, N., Collins-Atkins, C., Macpherson, R., Ford, S., and Palmer, A. (2006). Identifying the needs of carers in mental health services. *Nurs Times*, 102(17): 32–5.

Greppi, C. (2008). About evolutionary theories and geography: the missed encounter between Humboldt and Darwin. *Quaderni Storici*, 43(1): 33.

Gross, R., Sasson, Y., Zarhy, M., Zohar, J. (1998). Hospital environment in psychiatric hospital design. *General Hospital Psychiatry*, 20, 108–114.

Gunnell, D., Middleton, N., Whitley, E., Dorling, D., and Frankel, S. (2003). Influence of cohort effects on patterns of suicide in England and Wales, 1950–1999. *British Journal of Psychiatry*, 182, 164–70.

Gunnell, D., Middleton, N., Whitley, E., Dorling, D., and Frankel, S. (2003). Why are suicide rates rising in young men but falling in the elderly? A time-series analysis of trends in England and Wales 1950–1998. *Social Science and Medicine*, 57(4): PII S0277–9536(0202)00408–00402.

Hajat, S., Ebi, et al. (2003). The human health consequences of flooding in Europe and the implications for public health: a review of the evidence. *Applied Environmental Science and Public Health*, 1(1): 13–21.

Hall, E. (2000). Blood, brain and bone. *Area*, 32(1): 21–9.

Halpern, D. (1993). Minorities and Mental-health. *Social Science and Medicine*, 36(5): 597–607.

Halpern, D., and Nazroo, J. (2000). The ethnic density effect: Results from a national community survey of England and Wales. *International Journal of Social Psychiatry*, 46(1): 34–46.

Haney, T.J. (2007). "Broken windows" and Self-Esteem: Subjective understandings of neighborhood poverty and disorder. *Social Science Research*, 36(3): 968–94.

Hanifan, L. (1916). The rural school community center *Annals of the American Academy of Political and Social Science*, 67, 130–38.

Hanlon, P., and Carlisle, S. (2008). What can the science of well-being tell the discipline of psychiatrry – and why might psychiatry listen? *Advances in Psychiatric Treatment (2008)*, 14, 312–19.

Hansen, A., Bi, P., Nitschke, M., Ryan, P., Pisaniello, D., and Tucker, G. (2008). The effect of heat waves on mental health in a temperate Australian city. *Environmental Health Perspectives*, 116(10): 1369–75.

Hansen, E., and Donohoe, M. (2003). Health issues of migrant and seasonal farmworkers. *Journal of Health Care for the Poor and Underserved*, 14(2):153–64.

Harden, J. (2005). "Uncharted waters": The experience of parents of young people with mental health problems. *Qualitative Health Research*, 15(2): 207–23.

Harpham, T., Grant, E., and Rodriguez, C. (2004). Mental health and social capital in Cali, Colombia. *Social Science and Medicine*, 58(11): 2267–77.

Harrington, A. (ed.) (1997). *The Placebo Effect: An Interdisciplinary Exploration.* Cambridge, MA: Harvard University Press.

Harvey, D. (1989). *The Condition of Postmodernity*. Oxford: Blackwell.

Haslam, S.A., Jetten, J., Postmes, T., and Haslam, C. (2009). Social Identity, Health and Well-Being: An Emerging Agenda for Applied Psychology. *Applied Psychology – An International Review – Psychologie Appliquee – Revue Internationale*, 58(1): 1–23.

Havenaar, J. (1996). After Chernobyl. Research of psychological factors that affect health after a nuclear disaster [in Russian]. *Gos Nauchnyy Centr Socialnoy i Sudebnoy Psihiatrii im V.P. Serbskogo*.

Haynes, R.,and Bentham, G. (1979). *Community Hospitals and Rural Accessibility.* London: Croom Helm.

Henderson, C., Roux, A.V.D., Jacobs, D.R., Kiefe, C.I., West, D., and Williams, D.R. (2005). Neighbourhood characteristics, individual level socioeconomic factors, and depressive symptoms in young adults: the CARDIA study. *Journal of Epidemiology and Community Health*, 59(4): 322–8.

Herzog, T. (1985). A cognitive analysis of preference for waterscapes. *Journal of Environmental Psychology*, 5, 225–41.

Herzog, T., and Bosley, P. (1992). Tranquility and preference as affective Qualities of natural environments. *Journal of Environmental Psychology*, 12, 11–117.

Herzog, T., and Barnes G. (1999). Tranquility and preference revisited Journal of Environmental Psychology. *Journal of Environmental Psychology*, 19, 171–81.

Herzog, T., Chen, H., and Primeau, J. (2002). Perception of the restorative potential of natural and other settings. *Journal of Environmental Psychology*, 22, 295–306.

Hetherington, J. (1993). Is motion more important than it sounds?: The medium of presentation in environment perception research. *Journal of Environmental Psychology*, 13, 283–91.

Higginbotham, N., Connor, L., Albrecht, G., Freeman, S., and Agho, K. (2006). Validation of an environmental distress scale. *Ecohealth*, 3(4): 245–54.

Hobfoll, S.E., Freedy, J., Lane, C., and Geller, P. (1990). Conservation of Social Resources – Social Support Resource Theory. *Journal of Social and Personal Relationships*, 7(4): 465–78.

Hockey, J., Penhale, B., and Sibley, D. (2005). Environments of memory: home space, later life and grief, in J. Davidson, L. Bondi and M. Smith (eds), *Emotional Geographies*. Aldershot: Ashgate, pp. 135–45.

Hoek, H.W., Susser, E., Buck, K.A., Lumey, L.H., Lin, S.P., and Gorman, J. M. (1996). Schizoid personality disorder after prenatal exposure to famine. *American Journal of Psychiatry*, 153(12): 1637–39.

Holloway, J. (2006). Enchanted spaces: The seance, affect, and geographies of religion. *Annals of the Association of American Geographers*, 96(1): 182–7.

Holloway, S., and Valentine, G. (eds) (2000). *Children's Geographies.* London: Routledge.

Hothi, M., Bacon, N., Brophy, M., Mulgan, G. (2009). Neighbourliness + Empowerment = Wellbeing: Is there a formula for happy communities? 18 Victoria Park Square, Bethnal Green, London E2 9PF, United Kingdom: Young Foundation.

Hoyez, A.C. (2007). The 'world of yoga': The production and reproduction of therapeutic landscapes. *Social Science and Medicine*, 65(1): 112–24.

Hu, T.W. (2006). An international review of the national cost estimates of mental illness, 1990–2003. *Journal of Mental Health Policy and Economics*, 9(1): 3–13.

Hudson, R. (2004). Conceptualising Economies and their Geographies: Spaces, Flows and Circuits. *Progress in Human Geography*, 28(4): 447–71.

Hunt, M.O., Wise, L.R., Jipguep, M.C., Cozier, Y.C., and Rosenberg, L. (2007). Neighborhood racial composition and perceptions of racial discrimination: Evidence from the black women's health study. *Social Psychology Quarterly*, 70(3): 272–89.

Hunter, J.M., Shannon, G.W., and Sambrook, S.L. (1986). Rings Of Madness – Service Areas of 19th-century Asylums in North America. *Social Science and Medicine*, 23(10): 1033–50.

Hunter, J.S. (1985). Jarvis revisited: distance decay in service areas of 19th century asylums. *Professional Geographer*, 37(3): 296–203.

Huotari, A., and Herzig, K.H. (2008). Vitamin D and living in northern latitudes – An endemic risk area for vitamin D deficiency. *International Journal of Circumpolar Health*, 67(2–3): 164–78.

Huxley, P., Evans, S., Leese, M., Gately, C., Rogers, A., Thomas, R., et al. (2004). Urban regeneration and mental health. *Social Psychiatry and Psychiatric Epidemiology*, 39(4): 280–5.

Hwangbo, A. (1999). A new millennium and feng shui *Journal of Architecture*, 4(2): 191–8.

Hyndman, S.J. (1990). Housing Dampness and Health Amongst British Bengalis in East London. *Social Science and Medicine*, 30(1): 131–41.

IASC. (2007). Inter-Agency Standing Committee *Guidelines on mental health and psycho-social support in emergency settings*

Ingold, T. (2006). Rethinking the animate, re-animating thought. *Ethnos*, 71(1): 9–20.

Islam, M.K., Merlo, J., Kawachi, I., Lindstrom, M., Burstrom, K., and Gerdtham, U.-G. (2006). Does it really matter where you live? A panel data multilevel analysis of Swedish municipality-level social capital on individual health-related quality of life. *Health Econ Policy Law*, 1(3): 209–35.

Jablensky, A. (2000). Prevalence and incidence of schizophrenia spectrum disorders: implications for prevention. *Australian and New Zealand Journal of Psychiatry*, 34, S26–S34.

Jackson, L., Langille, L., Lyons, R., Hughes, J., Martin, D., and Winstanley, V. (2009). Does moving from a high-poverty to lower-poverty neighborhood improve mental health? A realist review of 'Moving to Opportunity'. *Health and Place*, 15(4): 961–70.

Janca, A. (2000). Telepsychiatry: an update on technology and its applications. *Current Opinion in Psychiatry*, 13(6): 591–97.

Jarman, B., and Hirsch, S. (1992). Statistical models to predict district psychiatric morbidity, in G. Thornicroft, C. Brewin and J. Wing (eds) *Measuring Mental Health Needs.* London: Gaskell (Royal College of Psychiatrists), pp. 62–80.

Jarvis, E. (1866). Influence of distance from and nearness to an insane hospital on its use by the people. *American Journal of Insanity*, 22, 361–418.

Jenkins, R., Bhugra, D., Bebbington, P., Brugha, T., Farrell, M., Coid, J., et al. (2008). Debt, income and mental disorder in the general population. *Psychological Medicine*, 38(10): 1485–93.

Jeon, Y.H., Brodaty, H., and Chesterson, J. (2005). Respite care for caregivers and people with severe mental illness: literature review. *Journal of Advanced Nursing*, 49(3): 297–306.

Jones, J. (2000). Mental health care reforms in Britain and Italy, since 1950: A cross-national comparative study. *Health and Place*, 6(3): 171–88.

Jones, K. and Moon, G. (1987). *Health, Disease and Society: An Introduction to Medical Geography*. London: Routledge.

Jones, O. and Cloke, P. (2002). *Tree Cultures: the Place of Trees and Trees in their Place*. Oxford: Berg.

Joseph, A., and Phillips, D. (1984). *Accessibility and Utilization: Geographical Perspectives on Health Care Delivery*. London: Harper Row.

Joseph, A.E., Kearns, R.A., and Moon, G. (2009). Recycling former psychiatric hospitals in New Zealand: Echoes of deinstitutionalisation and restructuring. *Health and Place*, 15(1): 79–87.

Joseph, A.E., and Moon, G. (2002). From retreat to health centre: legislation, commercial opportunity and the repositioning of a Victorian private asylum. *Social Science and Medicine*, 55(12): 2193–2200.

Jowett, B. (2009). *The Republic* By Plato Written 360 B.C.E

Judkins, G., Smith, M., and Keys, E. (2008). Determinism within human-environment research and the rediscovery of environmental causation. *Geographical Journal*, 174, 17–29.

Kahn, P. (1997). Development Psychology and the Biophilia Hypothesis: Children's Affiliation with Nature. *Development Review*, 17, 1–61.

Kahn, P., Severson, R, et al. (2009). The human relation with nature and technological nature. *Current Directions in Psychological Science*, 18(1): 37–42.

Kahn, R.S., Wise, P.H., Kennedy, B.P., and Kawachi, I. (2000). State income inequality, household income, and maternal mental and physical health: cross sectional national survey. *British Medical Journal*, 321(72): 1311–15.

Kammann, R., and Flett, R. (1983). A scale to measure current level of general happiness. *Australasian Psychologist*, 35, 259–65.

Kaplan, R.K. (ed.) (1989). *The Experience of Nature. A Psychological Perspective.* Cambridge: Cambridge University Press.

Kaplan, S. (1995). The Restorative Benefits of Nature – Toward an Integrative Framework. *Journal of Environmental Psychology*, 15(3): 169–82.

Kaplan, S., and Kaplan, R. (2003). Health, supportive environments, and the reasonable person model. *American Journal of Public Health*, 93(9): 1484–89.

Karlsen, S., and Nazroo, J. (2000). Identity and structure: rethinking inequalities and health, in H. Graham (ed.), *Understanding Health Inequalities*. Buckingham: Open University Press, pp. 38–57.

Kasof, J. (2009). Cultural variation in seasonal depression: Cross-national differences in winter versus summer patterns of seasonal affective disorder. *Journal of Affective Disorders*, 115(1–2): 79–86.

Kawachi, I., Kennedy, B., and Wilkinson, R. (1999). *Income Inequality and Health*, Volume 1. New York: New Press.

Kawachi, I., and Berkman, L.F. (2001). Social ties and mental health. *Journal of Urban Health – Bulletin of the New York Academy of Medicine*, 78(3): 458–67.

Kearney, A., and Bradley, J.J. (2009). 'Too strong to ever not be there': place names and emotional geographies. *Social and Cultural Geography*, 10(1): 77–94.

Kearns, R. (1993). Place and Health – Towards a Reformed Medical Geography. *Professional Geographer*, 45(2): 139–47.

Kearns, R., and Joseph, A. (1993). Space in its Place – Developing the Link in Medical Geography. *Social Science and Medicine*, 37(6): 711–17.

Kearns, R.A., and Joseph, A.E. (2000). Contracting opportunities: interpreting post-asylum geographies of mental health care in Auckland, New Zealand. *Health and Place*, 6(3): 159–69.

Kegel, M., Dam, H., Ali, F., and Bjerregaard, P. (2009). The prevalence of seasonal affective disorder (SAD) in Greenland is related to latitude. *Nordic Journal of Psychiatry*, 63(4): 331–5.

Keighren, I.M. (2006). Bringing geography to the book: charting the reception of Influences of geographic environment. *Transactions of the Institute of British Geographers*, 31(4): 525–40.

Kelling, G., and Coles, K. (1996). *Fixing Broken Windows: Restoring Order and Reducing Crime in Our Communities*. New York: Touchstone.

Kemm, J., Parry, J., and Palmer, S. (eds) (2004). *Health Impact Assessment: Concepts, Theory, Techniques and Applications*. Oxford: Oxford University Press.

Kendler, K.S. (2001). A psychiatric dialogue on the mind-body problem. *American Journal of Psychiatry*, 158(7): 989–1000.

Kendler, K.S. (2005). Toward a philosophical structure for psychiatry. *American Journal of Psychiatry*, 162(3): 433–40.

Kessler, R., Berglund, P., Glantz, M., Koretz, D., Merikangas, K., Walter, E., and Zawlavsky, A. (2004). Estimating the prevalence and correlates of serious mental illness in community epidemiological surveys In R. Manderscheid, Henderson, M. (eds), *Mental Health*, United States Center for Mental Health Services, Washington DC: US Govt. Printing Office, pp. 155–64.

Kessler, R.C., Galea, S., Jones, R.T., Parker, H.A. (2006). Mental illness and suicidality after Hurricane Katrina. *Bulletin of the World Health Organization*, 84(12): 930–9.

Kessler, R.C., and Wittchen, H.U. (2008). Post-disaster mental health need assessment surveys – the challenge of improved future research. *International Journal of Methods in Psychiatric Research*, 17, S1–S5.

Khawaja, M., Abdulrahim, S., Soweid, R.A.A., and Karam, D. (2006). Distrust, social fragmentation and adolescents' health in the outer city: Beirut and beyond. *Social Science and Medicine*, 63(5): 1304–15.

Khawaja, M., and Mowafi, M. (2006). Cultural capital and self-rated health in low income women evidence from the urban health study, Beirut, Lebanon. *Journal of Urban Health – Bulletin of the New York Academy of Medicine*, 83(3): 444–58.

Kienzler, H. (2008). Debating war-trauma and post-traumatic stress disorder (PTSD) in an interdisciplinary arena. *Social Science and Medicine*, 67(2): 218–27.

Kim, D., and Kawachi, I. (2007). US state-level social capital and health-related quality of life: Multilevel evidence of main, mediating, and modifying effects. *Annals of Epidemiology*, 17(4): 258–69.

Kingsley, J., Townsend, M., Phillips, R., Aldous, D. (2009). "If the land is healthy... it makes the people healthy": The relationship between caring for Country and health for the Yorta Yorta Nation, Boonwurrung and Bangerang Tribes. *Health and Place* 15(1): 291–9.

Kirkbride, J.B., Morgan, C., Fearon, P., Dazzan, P., Murray, R.M., and Jones, P.B. (2007). Neighbourhood-level effects on psychoses: re-examining the role of context. *Psychological Medicine*, 37(10): 1413–25.

Knapp, R. (1999). China's living houses: Folk beliefs, symbols, and household ornamentation. Honolulu: University of Hawai'i Press.

Knoblauch, H., Flick U., Maeder, C. with Lang, I. (2005). The State of the Art of Qualitative Research in Europe, *Forum for Qualitative Social Research*, 6, 3 – September 2005.

Knowles, C. (2000). Burger King, Dunkin' Donuts and community mental health care. *Health & Place*, 6(3): 213–24.

Kohen, D.E., Dahinten, V.S., Leventhal, T., and McIntosh, C.N. (2008). Neighborhood disadvantage: Pathways of effects for young children. *Child Development*, 79(1): 156–69.

Kong, L. (1990). Geography and Religion – Trends and Prospects. *Progress in Human Geography*, 14(3): 355–71.

Kong, L. (2001). Mapping 'new' geographies of religion: politics and poetics in modernity. *Progress in Human Geography*, 25(2): 211–33.

Koole, S. and Van den Berg, A. (2005). Lost in the Wilderness. *Journal of Personality and Social Psychology*, 88(5): 1014–28.

Koren, H.S., and Butler, C.D. (2006). The interconnection between the built environment ecology and health. *Environmental Security and Environmental Management: The Role of Risk Assessment*, 5, 111–25.

Kusenback, M. (2003). Street phenomenology: the go along as ethnographic research tool. *Ethnography*, 4(3): 455–85.

Kwan, M.P. (2004). GIS methods in time-geographic research: geo-computation and geovisualization of human activity patterns. *Geografiska Annaler (Series Brs*, 86(4): 267–80.

La Torre, M.A. (2006). Creating a healing environment. *Perspectives in Psychiatric Care*, 42(4): 262–4.

LaGreca, A., Silverman, W., Vernberg, E., Prinstein, M.J. (1996). Symptoms of posttraumatic stress in children after hurricane Andrew: A prospective study. *Journal of Consulting and Clinical Psychology*, 64(4): 712–23.

Lash, S., Szerszynski, B., Wynne, B. (1996). *Risk, Environment and Modernity: Towards a New Ecology*. London: Sage.

Latour, B. (1996). On actor-network theory – A few clarifications. *Soziale Welt-Zeitschrift Fur Sozialwissenschaftliche Forschung Und Praxis*, 47(4): 369.

Lawson, B., Phiri, M., Wells-Thorpe, J. (2003). *The Architectural Healthcare Environment and its Effect on Patient Health Outcomes*. London: HMSO

Lawson, V. (2007). Geographies of care and responsibility. *Annals of the Association of American Geographers*, 97(1): 1–11.

Lea, J. (2008). Retreating to nature: rethinking 'therapeutic landscapes'. *Area*, 40(1): 90–8.

Learmonth, A. (1988). *Disease Ecology*. Oxford: Blackwell.

Leventhal, T., and Brooks-Gunn, J. (2003). Moving to opportunity: an experimental study of neighborhood effects on mental health. *American Journal of Public Health*, 93(9): 1576–82.

Lewis, G., Bebbington, P., Brugha, T., Farrell, M., Gill, B., Jenkins, R., et al. (2003). Socio-economic status, standard of living, and neurotic disorder (Reprinted from *The Lancet*, vol. 352, pp. 605–609, 1998). *International Review of Psychiatry*, 15(1–2):;Lingler, J.H., Sherwood, P.R., Crighton, M.H., Song, M.K., and Happ, M.B. (2008). Conceptual challenges in the study of caregiver – Care recipient relationships. *Nursing Research*, 57(5): 367–72.

Little, J., and Leyshon, M. (2003). Embodied rural geographies: developing research agendas. *Progress in Human Geography*, 27(3): 257–72.

Little, J., Panelli, R., and Kraack, A. (2005). Women's fear of crime: A rural perspective. *Journal of Rural Studies*, 21(2): 151–63.

Little, J. and Leyshon, M. (2003). Embodied rural geographies: developing research agendas. *Progress in Human Geography*, 27(3): 257–72.

Lix, L.M., DeVerteuil, G., Walker, J.R., Robinson, J.R., Hinds, A.M., and Roos, L.L. (2007). Residential mobility of individuals with diagnosed schizophrenia – A comparison of single and multiple movers. *Social Psychiatry and Psychiatric Epidemiology*, 42(3): 221–8.

Longhurst, R. (2001). *Bodies: Exploring Fluid Boundaries*. London: Routledge.

Lorant, V., Deliege, D., Eaton, W., Robert, A., Philippot, P., and Ansseau, M. (2003). Socioeconomic inequalities in depression: A meta-analysis. *American Journal of Epidemiology*, 157(2): 98–112.

Loukaki, A. (1997). Whose genius loci?: contrasting interpretations of the "sacred rock of the Athenian Acropolis". *Annals of the Association of American Geographers*, 87(2): 306–29.

Luchins, A.S. (1989). Moral treatment in asylums and general hospitals in 19th-Century America. *Journal of Psychology*, 123(6): 585–607.

Luke, D.A., and Harris, J.K. (2007). Network analysis in public health: History, methods, and applications. *Annual Review of Public Health*, 28, 69–93.

Lupton, D. (1993). Risk as moral danger: the social and political functions of risk discourse in public health. *International Journal of Health Services*, 23(3): 425–35.

Lupton, D., and Petersen, A. (1996). *The New Public Health: Health and Self in the Age of Risk*. London: Sage.

Lupton, D. (1998). *The Emotional Self: A Sociocultural Exploration*. London: Sage.

Lupton, D. (1999). *Risk*. London: Routledge.

Maas, J., Verheij, R.A., Groenewegen, P.P., de Vries, S., and Spreeuwenberg, P. (2006). Green space, urbanity, and health: how strong is the relation? *Journal of Epidemiology and Community Health*, 60(7): 587–92.

Mair, C., Diez-Roux, A., Galea, S. (2008). Are neighbourhood characteristics associated with depressive symptoms? A review of evidence. *Journal of Epidemiology and Community Health*, 62(11): 940–U921.

Mak, M., and Ng, S.T. (2005). The art and science of Feng Shui – a study on architects' perception. *Building and Environment*, 40(3): 427–34.

Mallinson, S., and Popay, J. (2007). Describing depression: ethnicity and the use of somatic imagery in accounts of mental distress. *Sociology of Health and Illness*, 29, 857–71.

Mangalore, R., Knapp, M., and Jenkins, R. (2007). Income-related inequality in mental health in Britain: the concentration index approach. *Psychological Medicine*, 37(7): 1037–45.

Mann, G. (2008). A Negative Geography of Necessity. *Antipode*, 40(5): 991–34.

Mari, G. (2006). Unity, era and world in La 'Mediterranee'. *Rethinking History*, 10(1): 63–74.

Marmot, M., and Siegrist, J. (2004). Health inequalities and the psychosocial environment. *Social Science and Medicine*, 58(8): 1461–61.

Massey, D. (2005). *For Space*. London: Sage.

Matthews, I. (2007). *Body Subjects and Disordered Minds*. Oxford: Oxford University Press.

Matthis, I. (2000). Sketch for a metapsychology of affect. *International Journal of Psychoanalysis*, 81, 215–28.

Maylath, E., Seidel, J., Werner, B. et al. (1999). Geographical analysis of the risk of psychiatric hospitalization in Hamburg from 1988–1994. *European Psychiatry*, 14(8): 414–25.

McAllister, C.L., Thomas, T.L., Wilson, P.C., and Green, B.L. (2009). Root Shock Revisited: Perspectives of Early Head Start Mothers on Community and Policy Environments and Their Effects on Child Health, Development, and School Readiness. *American Journal of Public Health*, 99(2): 205–10.

McCrone, P., Thornicroft, G., Boyle, S., Knapp, M., and Aziz, F. (2006). The development of a Local Index of Need (LIN) and its use to explain variations in social services expenditure on mental health care in England. *Health and Social Care in the Community*, 14(3): 254–63.

McFarlane, A. (2008). Psychiatric morbidity following disasters, in H. Freeman and S. Stansfeld (eds), *The Impact of Environment on Psychiatric Disorder*. Hove, UK: Routledge, pp. 284–310.

McGinty, K.L., Saeed, S.A., Simmons, S.C., and Yidirim, Y. (2006). Telepsychiatry and e-mental health services: Potential for improving access to mental health care. *Psychiatric Quarterly*, 77(4): 335–42.

McGregor, J. (2008). Abject spaces, transnational calculations: Zimbabweans in Britain navigating work, class and the law *Transactions of the Institute of British Geographers*, 33(4): 466–82.

McGrew, W.C. (2007). New wine in new bottles – Prospects and pitfalls of cultural primatology. *Journal of Anthropological Research*, 63(2): 167–83.

McKenzie, K., Whitley, R., and Weich, S. (2002). Social capital and mental health. *British Journal of Psychiatry*, 181, 280–3.

Meade, M., and Earickson, R. (2000). *Medical Geography*. Second Edition. New York: Guilford Press.

Meltzer, H., Vostanis, P., Goodman, R., and Ford, T. (2007). Children's perceptions of neighbourhood trustworthiness and safety and their mental health. *Journal of Child Psychology and Psychiatry*, 48(12): 1208–13.

Merleau-Ponty, M. (2001). *Phenomenology of Perception*. 2nd edition London: Routledge.

Mezey, G., and Robbins, I. (2001). Usefulness and validity of post-traumatic stress disorder as a psychiatric category. *British Medical Journal*, 323(7312): 561–3.

Middleton, N., Sterne, J., and Gunnell, D. (2008). An atlas of suicide mortality: England and Wales, 1988–1994. *Health and Place*, 14(3): 492–506.

Middleton, N., and Gunnell, D. (2000). Trends in suicide in England and Wales. *British Journal of Psychiatry*, 176, 595.

Middleton, N., Gunnell, D., Frankel, S., Whitley, E., and Dorling, D. (2003). Urban–rural differences in suicide trends in young adults: England and Wales, 1981–1998. *Social Science and Medicine*, 57(7): 1183–94.

Middleton, N., Sterne, J. and Gunnell, D. (2006). The geography of despair among 15 44-year-old men in England and Wales: putting suicide on the map. *Journal of Epidemiology and Community Health*, 60(12): 1040–47.

Middleton, N., Whitley, E., Frankel, S., Dorling, D., Sterne, J., and Gunnell, D. (2004). Suicide risk in small areas in England and Wales, 1991–1993. *Social Psychiatry and Psychiatric Epidemiology*, 39(1): 45–52.

Milligan, C. (1996). Service dependent ghetto formation: a transferable concept? *Health and Place*, 2(4): 199–211.

Milligan, C. (1999). Without these walls: a geography of mental ill-health in a rural settlement, in R. Butler, and H. Parr (eds) *Mind and Body Spaces*. London: Routledge, pp. 221–39.

Milligan, C. (2000a). 'Breaking out of the asylum': developments in the geography of mental ill-health – the influence of the informal sector. *Health and Place*, 6(3): 189–200.

Milligan, C. (2000b). 'Bearing the burden': towards a restructured geography of caring. *Area*, 32(1): 49–58.

Milligan, C. (2005). From home to 'home': situating emotions within the caregiving experience. *Environment and Planning A*, 37(12): 2105–20.

Milligan, C., Bingley, A., Gattrell, A. (2005). Healing and Feeling: The Place of Emotion in Later Life, in J. Davidson, L. Bondi and M. Smith (eds), *Emotional Geographies*. Aldershot: Ashgate, pp. 49–62.

Milligan, C., and Bingley, A. (2007). Restorative or scary spaces? The impact of woodland on the mental well-being of young adults. *Health and Place*, 13(4): 799–811.

Milligan, C., Gatrell, A., and Bingley, A. (2004). 'Cultivating health': therapeutic landscapes and older people in northern England. *Social Science and Medicine*, 58(9): 1781–93.

Milton, J. (1647). *Paradise Lost*. Electronic version available at http://en.wikisource.org/wiki/Paradise_Lost_(1674) accessed 29.11.09.

Mindell, J., Boaz, A., Joffe, M., Curtis, S., and Birley, M. (2004). Enhancing the evidence base for health impact assessment. *Journal of Epidemiology and Community Health*, 58(7): 546–51.

Mindell, J., Ison, E., and Joffe, M. (2003). A glossary for health impact assessment. *Journal of Epidemiology and Community Health*, 57(9): 647–51.

Mitchell, K. (2004). Geographies of identity: multiculturalism unplugged. *Progress in Human Geography*, 28(5): 641–51.

Mitchell, K. (2006). Geographies of identity: the new exceptionalism. *Progress in Human Geography*, 30(1): 95–106.

Mitchell, K. (2007). Geographies of identity: the intimate cosmopolitan. *Progress in Human Geography*, 31, 706–20.

Mitchell, R., and Popham, F. (2008). Effect of exposure to natural environment on health inequalities: an observational population study. *Lancet*, 372(9650): 1655–60.

Moerman, D. (2002). *Meaning, Medicine and the 'Placebo Effect'*. Cambridge: Cambridge University Press.

Mohamud, S., Bhui, K., Craig, T., Warfa, N., Stansfeld, S., Curtis, S., et al. (2004). Residential mobility and mental health: A cross-sectional study of Somali refugees in London. *Ethnicity and Health*, 9, S80–S81.

Mohan, J. (2002). *Planning, Markets and Hospitals*. London: Routledge.

Moon, G. (2000). Risk and protection: the discourse of confinement is contemporary mental health policy. *Health and Place*, 6(3): 239–50.

Moon, G., Joseph, A.E., and Kearns, R. (2005). Towards a general explanation for the survival of the private asylum. *Environment and Planning C: Government and Policy*, 23(2): 159–72.

Moon, G., Kearns, R., and Joseph, A. (2006). Selling the private asylum: therapeutic landscapes and the (re)valorization of confinement in the era of community care. *Transactions of the Institute of British Geographers*, 31(2): 131–49.

Moore, N., and Whelan, Y. (eds) (2007). *Heritage, Memory and the Politics of Identity: New Perspectives on the Cultural Landscape*. Aldershot: Ashgate.

Moore, N.C., and Nelson, W.H. (1998). Five papers on psychiatry in rural areas: An introduction. *Psychiatric Services*, 49(7): 957.

Moos, R.H. (1997). The social climate of hospital programs, in R.H.E. Moos (ed.), *Evaluating Treatment Environments: The Quality of Psychiatric and Substance Abuse Programs*, 2nd Edition. New Brunswick, NJ: Transaction Publishers. pp. 23–44.

Morgan, C., and Fearon, P. (2007). Social experience and psychosis – Insights from studies of migrant and ethnic minority groups. *Epidemiologia E Psichiatria Sociale – An International Journal for Epidemiology and Psychiatric Sciences*, 16(2): 118–23.

Morita, E., Fukuda, S., Nagano, J., Hamajima, N., Yamamoto, H., Iwai, Y., et al. (2007). Psychological effects of forest environments on healthy adults: Shinrin-yoku (forest-air bathing, walking) as a possible method of stress reduction. *Public Health*, 121(1): 54–63.

Morley, D. (2000). *Home Territories: Medial, Mobility and Identity*. London: Routledge.

Morris, D. (2004). *The Nature of Happiness*. London: Little Books Ltd.

Morris, R., and Carstairs, V. (1991). Which deprivation? – a comparison of selected deprivation indices. *Journal of Public Health Medicine*, 13(4): 318–26.

Morrow, V. (1999). Conceptualising social capital in relation to the well-being of children and young people: a critical review. *Sociological Review*, 47(4): 744–65.

Morrow, V. (2004). Networks and neighbourhoods: children's accounts of friendships, family and place, in C. Phillipson, A. Graham, and D. Morgan (eds),

Social Networks and Social Exclusion: Sociological and Policy Perspectives. Aldershot: Ashgate.

Moser, G. (1984). Water quality perception, a dynamic evaluation. *Journal of Environmental Psychology*, 4 (201–10).

Muramatsu, N. (2003). County-level income inequality and depression among older Americans. *Health Services Research*, 38(6): 1863–83.

Murdoch, J. (1997). Inhuman/nonhuman/human: actor-network theory and the prospects for a nondualistic and symmetrical perspective on nature and society. *Environment and Planning D: Society and Space*, 15(6): 731–56.

Murdoch, J. (1998). The spaces of actor network theory. *Geoforum*, 29(4): 357–74.

Murdoch, J., Marsden, T., and Banks, J. (2000). Quality, nature, and embeddedness: Some theoretical considerations in the context of the food sector. *Economic Geography*, 76(2): 107–25.

Murray, C., and Lopez, A. (1997). Global mortality, disability, and the contribution of risk factors: Global Burden of Disease Study. *Lancet*, 17(349(9063)): 1436–42.

Myers, G.J., Davidson, P.W., Weitzman, M., and Lanphear, B.P. (1997). Contribution of heavy metals to developmental disabilities in children. *Mental Retardation and Developmental Disabilities Research Reviews*, 3(3): 239–45.

Nancy, J.-L. (2000). Being singular-plural, in R. Richardson and A. O'Byrne (translators) (ed.). Stanford, US: Stanford University Press.

Nandi, A., Galea, S., Ahern, J., Bucciarelli, A., Vlahov, D., and Tardiff, K. (2006). What explains the association between neighborhood-level income inequality and the risk of fatal overdose in New York City? *Social Science and Medicine*, 63(3): 662–74.

Napreyenko A., and Logahanovsky K. (1995). Systematizing psychological disorders connected to the consequences of the Chernobyl accident *Vrach Delo*, 5–6, 25–8.

Nast, H., and Pile, S. (eds) (1998). *Places through the Body*. London: Routledge.

Neeleman, J., and Wessely, S. (1999). Ethnic minority suicide: a small area geographical study in south London. *Psychological Medicine*, 29(2): 429–36.

Neeleman, J., Wilson-Jones, C., and Wessely, S. (2001). Ethnic density and deliberate self harm; a small area study in south east London. *Journal of Epidemiology and Community Health*, 55(2): 85–90.

Neria, Y., Nandi, A., and Galea, S. (2008). Post-traumatic stress disorder following disasters: a systematic review *Psychological Medicine*, 38(4): 467–80.

Neugebauer, R. (2005). Schizophrenia and the Chinese famine of 1959–1961 – Reply. *Jama – Journal of the American Medical Association*, 294(23): 2968–9.

Neugebauer, R., Hoek, H.W., and Susser, E. (1999). Prenatal exposure to wartime famine and development of antisocial personality disorder in early adulthood. *Jama – Journal of the American Medical Association*, 282(5): 455–62.

Nicolini, D. (2006). The work to make telemedicine work: A social and articulative view. *Social Science and Medicine*, 62(11): 2754–67.

Nicolson, P., and Wilson, R. (2004). Is domestic violence a gender issue? Views from a British City. *Journal of Community and Applied Social Psychology*, 14(4): 266–83.

Nielsen, E.H., Simonsen, K. (2003). Scaling from 'below': Practices, strategies and urban spaces. *European Planning Studies*, 11(8): 911–27.

Noble, M., Wright, G., Dibben, C., Smith, G.A.N., McLennan, D., Anttila, C., Barnes, H., Mokhtar, C., Noble, S., Avenell, D., Gardner, J., Covizzi, I., Lloyd, M. (2004). *The English Indices of Deprivation 2004*. W. Yorkshire: ODPM Publications.

Nogue, J., and Vicente, J. (2004). Landscape and national identity in Catalonia. *Political Geography*, 23(2): 113–32.

Nolh, S., Kaspar, V., and Wickrama, K.A.S. (2007). Overt and subtle racial discrimination and mental health: Preliminary findings for Korean immigrants. *American Journal of Public Health*, 97(7): 1269–74.

Norman, S. (2006). The use of telemedicine in psychiatry. *Journal of Psychiatric and Mental Health Nursing*, 13(6): 771–7.

Norris, F., and Hamblen, J. (2004). Standardized self-report measures of civilian trauma and PTSD, in T. Keane and T. Martin (eds), *Assessing Psychological Trauma and PTSD*. New York: Guilford Press, pp. 63–102.

Norris, F. (2005). Range, magnitude and duration of the effects of disasters on mental health. Review update 2005. Dartmouth Medical School and National Centre for PTSD. Retrieved 23.06.09 from http://katasztrofa.hu/documents/Research_Education_Disaster_Mental_Health.pdf.

Norris, F.H., Friedman, M.J., Watson, P.J., Byrne, C.M., Diaz, E., and Kaniasty, K. (2002a). 60,000 disaster victims speak: Part I. An empirical review of the empirical literature, 1981–2001. *Psychiatry – Interpersonal and Biological Processes*, 65(3): 207–39.

Norris, F.H., Friedman, M.J., and Watson, P.J. (2002b). 60,000 disaster victims speak: Part II. Summary and implications of the disaster mental health research. *Psychiatry – Interpersonal and Biological Processes*, 65(3): 240–60.

Norris, F.H., and Kaniasty, K. (1996). Received and perceived social support in times of stress: A test of the social support deterioration deterrence model. *Journal of Personality and Social Psychology*, 71(3): 498–511.

Oakes, J.M. (2004). The (mis)estimation of neighbourhood effects: Causal inference for a practicable epidemiology. *Social Science and Medicine*, 58, 1929–52.

Oakes, J.M. (2009). Commentary: Individual, ecological and multilevel fallacies. *International Journal of Epidemiology*, 38(2): 361–68.

Oberhelman, S.M. (1981). The Interpretation of Prescriptive Dreams in Ancient Greek Medicine. *Journal of the History of Medicine and Allied Sciences*, 36(4): 416–24.

Obunseitan, O. (2005). Topophilia and the quality of Life. *Environmental Health Perspectives*, 113(2): 143–8.

O'Byrne, P., and Holmes, D. (2007). The micro-fascism of Plato's good citizen: producing (dis)order through the construction of risk. *Nurs Philos*, 8(2): 92–101.

O'Campo, P., Salmon, C., and Burke, J. (2009). Neighbourhoods and mental well-being: What are the pathways? *Health and Place*, 15(1): 56–68.

Olwig, K.R. (1996). Recovering the substantive nature of landscape. *Annals of the Association of American Geographers*, 86(4): 630–53.

Ornelas, I.J., Amell, J., Tran, A.N., Royster, M., Armstrong-Brown, J., and Eng, E. (2009). Understanding African American Men's Perceptions of Racism, Male Gender Socialization, and Social Capital Through Photovoice. *Qualitative Health Research*, 19(4): 552–65.

Owen, G., and Harland, R. (2001). Editor's Introduction: Theme Issue on Phenomenlology and Psychiatry for the 21st Century. Taking Phenomenology Seriously. *Schizophrenia Bulletin*, 33(1): 105–107.

Pain, R. (1997). Old age and ageism in urban research: the case of fear. *International Journal of Urban and Regional Studies*, 21(1): 117–28.

Pain, R. (2004). Social geography: participatory research. *Progress in Human Geography*, 28(5): 652–63.

Pain, R. (2006). Paranoid parenting? Rematerializing risk and fear for children. *Social and Cultural Geography*, 7(2): 221–43.

Pain, R., Grundy, S., Gill, S., Towner, E., Sparks, G., and Hughes, K. (2005). 'So long as I take my mobile': Mobile phones, urban life and geographies of young people's safety. *International Journal of Urban and Regional Research*, 29(4): 814.

Pain, R., and Kindon, S. (2007). Participatory geographies. *Environment and Planning A*, 39(12): 2807–12.

Pain, R., MacFarlane, R., and Turner, K. (2006). 'When, where, if, and but': qualifying GIS and the effect of streetlighting on crime and fear. *Environment and Planning A*, 38(11): 2055–74.

Palka, E. (1999). Accessible wilderness as a therapeutic landscape: experiencing nature of Denali National Park, Alaska, in A. Williams (ed.), *Therapeutic Landscapes: The Dynamic between Place and Wellness*. New York: University Press of America, pp. 29–52.

Pampalon, R., Hamel, D., AND Raymond, G. (2004). Indice de défavorisation pour l'étude de la santé et du bien-être au Québec – Mise à jour 2001. Quebec City, Quebec, Canada: Institut National de Santé Publique au Québec.

Panelli, R. and Tipa, G. (2007). Placing well-being: a Maori case study of cultural and environmental specificity. *Ecohealth*, 4, 445–60.

Park, B.J., Tsunetsugu, Y., Ishii, H., Furuhashi, S., Hirano, H., Kagawa, T., et al. (2008). Physiological effects of Shinrin-yoku (taking in the atmosphere of the forest) in a mixed forest in Shinano Town, Japan. *Scandinavian Journal of Forest Research*, 23(3): 278–83.

Park, D., and Radford, J. (1997). Space, place and the asylum. *Health and Place*, 3(2): 71–2.

Park, R., Burgess, E., and McKenzie, R. (1925). *The City*. Chicago: University of Chicago Press.

Parr, H. (1997). Mental health, public space, and the city: Questions of individual and collective access. *Environment and Planning D: Society and Space*, 15(4): 435–54.

Parr, H. (1999). Delusional geographies: the experiential worlds of people during madness/illness. *Environment and Planning D: Society and Space*, 17(6): 673–90.

Parr, H. (2000). Interpreting the 'hidden social geographics' of mental health: ethnographies of inclusion and exclusion in semi-institutional places. *Health and Place*, 6(3): 225–37.

Parr, H. (2002). New body-geographies: the embodied spaces of health and medical information on the Internet. *Environment and Planning D: Society and Space*, 20(1): 73–95.

Parr, H. (2002a). Diagnosing the body in medical and health geography 1999–2000. *Progress in Human Geography*, 26, 240–51.

Parr, H. (2002b). New body-geographies: the embodied spaces of health and illness information on the internet. *Environment and Planning D: Society and Space*, 20, 73–95.

Parr, H. (2003). Medical geography: care and caring. *Progress in Human Geography*, 27(2): 212–21.

Parr, H. (2004). Medical geography: critical medical and health geography? *Progress in Human Geography*, 28(2): 246–57.

Parr, H. (2006). Mental health, the arts and belongings. *Transactions of the Institute of British Geographers*, 31(2): 150–66.

Parr, H. (2007). Mental health, nature work, and social inclusion. *Environment and Planning D: Society and Space*, 25(3): 537–61.

Parr, H., and Philo, C. (2003). Rural mental health and social geographies of caring. *Social and Cultural Geography*, 4(4): 471–88.

Parr, H., Philo, C., and Burns, N. (2003). 'that awful place was home': Reflections on the contested meanings of Craig Dunain asylum. *Scottish Geographical Journal*, 119(4): 341–60.

Parr, H., Philo, C., and Burns, N. (2004). Social geographies of rural mental health: experiencing inclusions and exclusions. *Transactions of the Institute of British Geographers*, 29(4): 401–19.

Parry, G., Van Cleemput, P., Peters, J., Walters, S., Thomas, K., and Cooper, C. (2007). Health status of Gypsies and Travellers in England. *Journal of Epidemiology and Community Health*, 61(3): 198–204.

Patel, V., Araya, R., de Lima, M., Ludermir, A., and Todd, C. (1999). Women, poverty and common mental disorders in four restructuring societies. *Social Science and Medicine*, 49(11): 1461–71.

Patel, V., and Kleinman, A. (2003). Poverty and common mental disorders in developing countries. *Bulletin of the World Health Organization*, 81(8): 609–15.

Pederson, A., Lupton, D. (1996). *The New Public Health: Health and Self in the Age of Risk*. London: Sage.

Peek, F. (1994). *The Real Rain Man: Kim Peek*. Salt Lake City: Harkness.

Peet, R. (1993). Geographical Determinism in Fin-de-siècle Marxism – Plekhanov, Georgii and the Environmental Basis of Russian History – Reinventing Marxist Geography – A Critique Of Bassin. *Annals of the Association of American Geographers*, 83(1): 156–60.

Penning-Rowsell, E., and Lowenthal, D. (eds) (1986). *Landscape Meanings and Values*. London: Allen and Unwin.

Peterson, L.E., Tsai, A.C., Petterson, S., and Litaker, D.G. (2009). Rural–urban comparison of contextual associations with self-reported mental health status. *Health and Place*, 15(1): 125–32.

Phifer, J.F. (1990). Psychological Distress and Somatic Symptoms after Natural Disaster – Differential Vulnerability Among Older Adults. *Psychology and Aging*, 5(3): 412–20.

Phillipson, C., Allan, G., and Morgan, D. (eds.) (2004). *Social Networks and Social Exclusion: Social and Policy Perspectives*. Aldershot: Ashgate.

Philo, C. (1987). Not at our Seaside – Community Opposition to a 19th-century Branch Asylum. *Area*, 19(4): 297–302.

Philo, C. (1989). 'Enough to drive one mad': the organisation of space in nineteenth-century lunatic asylums, in J. Wolch, and M. Dear (eds), *The Power of Geography: How Territory Shapes Social Life*. London: Unwin Hyman, pp. 258–90.

Philo, C. (1992). Foucault Geography. *Environment and Planning D: Society and Space*, 10(2): 137–61.

Philo, C. (1995). Journey to asylum – A medical-geographical idea in Historical context. *Journal of Historical Geography*, 21(2): 148–68.

Philo, C. (1997). Across the water: reviewing geographical studies of asylums and other mental health facilities. *Health Place*, 3(2): 73–89.

Philo, C., and Wilbert, C. (2000). *Animal Spaces, Beastly Places: New Geographies of Human–Animal Relations*. London: Routledge.

Philo, C. (2000). *The Birth of the Clinic*: an unknown work of medical geography. *Area*, 32(1): 11–19.

Philo, C. (2000). Post-asylum geographies: an introduction. *Health and Place*, 6(3): 135–36.

Philo, C., Parr, H., and Burns, N. (2003). Rural madness: a geographical reading and critique of the rural mental health literature. *Journal of Rural Studies*, 19(3): 259–81.

Philo, C. (2004). *A Geographical History of Institutional Provision for the Insane from Medieval Times to the 1860s*. Lewiston, NY: Edwin Mellen Press.

Philo, C. (2005). The geography of mental health: an established field? *Current Opinion in Psychiatry*, 18(5): 585–91.

Philo, C., Parr, H., and Burns, N. (2005). "An oasis for us": 'in-between' spaces of training for people with mental health problems in the Scottish Highlands. *Geoforum*, 36(6): 778–91.

Philo, C. (2006). Madness, memory, time, and space: the eminent psychological physician and the unnamed artist–patient. *Environment and Planning D: Society and Space*, 24(6): 891–917.

Philo, C., and Metzel, D.S. (2005). Introduction to theme section on geographies of intellectual disability: 'outside the participatory mainstream'? *Health and Place*, 11(2): 77–85.

Philo, C., and Parr, H. (2000). Institutional geographies: introductory remarks. *Geoforum*, 31(4): 513–21.

Philo, C., and Parr, H. (2003). Introducing psychoanalytic geographies. *Social and Cultural Geography*, 4(3): 283–93.

Philo, C., and Parr, H. (2004). 'They shut them out the road': Migration, Mental Health and the Scottish Highlands. *Scottish Geographical Journal*, 120(1–2): 47–70.

Philo, C., and Wolch, J. (2001). The 'three waves' of research in mental health geography: a review and critical commentary. *Epidemiol Psichiatr Soc*, 10(4): 230–44.

Piccardi, L., Monti, C., Vaselli, O., Tassi, F., Gaki-Papanastassiou, K., and Papanastassiou, D. (2008). Scent of a myth: tectonics, geochemistry and geomythology at Delphi (Greece). *Journal of the Geological Society*, 165, 5–18.

Pickett, K.E., James, O.W., and Wilkinson, R.G. (2006). Income inequality and the prevalence of mental illness: a preliminary international analysis. *Journal of Epidemiology and Community Health*, 60(7): 646–47.

Pickett, K.E., and Pearl, M. (2001). Multilevel analyses of neighbourhood socioeconomic context and health outcomes: a critical review. *Journal of Epidemiology and Community Health*, 55(2): 111–22.

Pickett, K.E., and Wilkinson, R.G. (2008). People like us: ethnic group density effects on health. *Ethnicity and Health*, 13(4): 321–34.

Picot, J. (1998). Telemedicine and telehealth in Canada: Forty years of change in the use of information and communications technologies in a publicly administered health care system. *Telemedicine Journal*, 4(3): 199–205.

Pinfold, V. (2000). 'Building up safe havens...all around the world': users experiences of living in the community with mental health problems. *Health and Place*, 6(3): 201–12.

Popay, J., Thomas, C., Williams, G., Bennett, S., Gatrell, A., and Bostock, L. (2003). A proper place to live: health inequalities, agency and the normative dimensions of space. *Social Science and Medicine*, 57(1): 55–69.

Popay, J., Kowarzik, U., Mallinson, S., Mackian, S., and Barker, J. (2007). Social problems, primary care and pathways to help and support: addressing health inequalities at the individual level. Part II: Lay perspectives. *Journal of Epidemiology and Community Health*, 61, 972–7.

Popay, J., Bennett, S., Thomas, C., Williams, G., Gatrell, A., and Bostock, L. (2003). Beyond 'beer, fags, egg and chips'? Exploring lay understandings of social inequalities in health. *Sociology of Health and Illness*, 25(1): 1–23.

Popke, E.J. (2003). Poststructuralist ethics: subjectivity, responsibility and the space of community. *Progress in Human Geography*, 27(3): 298–316.

Porteous, J. (1990). *Landscapes of the mind: Worlds of Sense and Metaphor*. Boston: University of Toronto Press.

Portes, A. (1998). Social capital: its origins and applications in modern sociology. *Annual Review of Sociology*, 24, 1–24.

Portes, A. (2000). The two meanings of social capital. *Sociological Forum*, 15(1): 1–12.

Power, A. (2008). Caring for independent lives: Geographies of caring for young adults with intellectual disabilities. *Social Science and Medicine*, 67(5): 834–43.

Probyn, E. (1996). *Outside Belongings*. New York: Routledge.

Proctor, J. (2006a). Introduction: Theorizing and studying religion. *Annals of the Association of American Geographers*, 96(1): 165–68.

Proctor, J. (2006b). Religion as trust in authority: Theocracy and ecology in the United States. *Annals of the Association of American Geographers*, 96(1): 188–96.

Prudhoe, T. L. (1984). Environmental Determinism and a Selection of Current Geographical Literature – A Functional Malthusian Link. *Ohio Journal of Science*, 84(2): 34–34.

Pulcino, T., Galea, S., Ahern, J., Resnick, H., Foley, M., and Vlahov, D. (2003). Posttraumatic stress in women after the September 11 terrorist attacks in New York City. *Journal of Women's Health*, 12(8): 809–20.

Putnam, R. (2000). *Bowling Alone: The Collapse and Revival of American Community*. New York: Simon & Schuster.

Quirk, A., Lelliott, P. and Seale, C. (2004). Service users' strategies for managing risk in the volatile environment of an acute psychiatric ward. *Social Science and Medicine*, 59, 2573–83.

Quirk, A., Lelliott, P., and Seale, C. (2006). The permeable institution: An ethnographic study of three acute psychiatric wards in London. *Social Science and Medicine*, 63(8): 2105–17.

Ramsay, R. (1990). Invited Review – Post-traumatic-stress disorder – A New Clinical Entity. *Journal of Psychosomatic Research*, 34(4): 355–65.

Rashid, S.F., and Michaud, S. (2000). Female adolescents and their sexuality: Notions of honour, shame, purity and pollution during the floods. *Disasters*, 24(1): 54–70.

Rawls, J. (1972). *A Theory of Justice*. Oxford: Clarendon Press.

Read, R. (2008). The 'hard' problem of consciousness is continually reproduced and made harder by all attempts to solve it. *Theory Culture and Society*, 25(2): 51–86.

Rehkopf, D., and Buka, S. (2006). The association between suicide and the socio-economic characteristics of geographical areas: a systematic review *Psycholological Medicine*, 36 (2): 145–57.

Rezaeian, M., Dunn, G., St Leger, S., and Appleby, L. (2005). The ecological association between suicide rates and indices of deprivation in English local authorities. *Social Psychiatry and Psychiatric Epidemiology*, 40(10): 785–91.

Richardson, B.C. (1996). Detrimental determinists: Applied environmentalism as bureaucratic self-interest in the Fin-de-Siècle British Caribbean. *Annals of the Association of American Geographers*, 86(2): 213–34.

Richmond, C., and Ross, N. (2009). The determinants of First Nation and Inuit health: a critical population health approach. *Health and Place*, 15, 403–11.

Richmond, C., Elliott, S.J., Matthews, R., and Elliott, B. (2005). The political ecology of health: perceptions of environment, economy, health and well-being among 'Namgis First Nation. *Health and Place*, 11(4): 349–65.

Riva, M., Gauvin, L, and Barnett, T. (2007). Towards the next generation of research into small area effects on health: a synthesis of multilevel investigations published since July 1998. *Journal of Epidemiology and Community Health*, 61, 853–62.

Riva, M., Curtis, S., Gauvin, L., and Fagg, J. (2009). Unravelling the extent of inequalities in health across urban and rural areas: evidence from a national sample in England. *Soc Sci Med*, 68(4): 654–63.

Rival, L. (1998). *The Social Life of Trees: Anthropological Perspectives on Tree Symbolism*. Oxford: Berg.

Robertson, C., Halcon, L., Savik, K., Johnson, D., Spring, M., Butcher, J., et al. (2006). Somali and Oromo refugee women: trauma and associated factors. *Journal of Advanced Nursing*, 56(6): 577–87.

Roick, C., Heider, D., Bebbington, P.E., Angermeyer, M.C., Azorin, J.M., Brugha, T.S., et al. (2007). Burden on caregivers of people with schizophrenia: comparison between Germany and Britain. *British Journal of Psychiatry*, 190, 333–38.

Rose, G. (2004). 'Everyone's cuddled up and it just looks really nice': an emotional geography of some mums and their family photos. *Social and Cultural Geography*, 5, 549–64.

Rumyantseva, G., Levina, T., Lebedeva, M., Chinkina, O., Melnichuk, T., Margolina, V., Pliplina, D., Sokolova, T., Grushkov, A. (1996). Research of psychological factors that affect health after a nuclear disaster [in Russian].

Ryan-Nicholls, K.D., and Haggarty, J.M. (2007). Collaborative mental health in rural and isolated Canada: stakeholder feedback. *J Psychosoc Nurs Ment Health Serv*, 45(12): 37–45.

Saegert, S.C., Klitzman, S., Freudenberg, N., Cooperman-Mroczek, J., and Nassar, S. (2003). Healthy housing: A structured review of published evaluations of US interventions to improve health by modifying housing in the United States, 1990–2001. *American Journal of Public Health*, 93(9): 1471–7.

Samet, J.M., and Spengler, J.D. (2003). Indoor environments and health: Moving into the 21st century. *American Journal of Public Health*, 93(9): 1489–93.

Sampson, R., Morenoff, J., and Gannon-Rowley, T. (2002). Assessing "neighborhood effects": Social processes and new directions in research. *Annual Review of Sociology*, 28, 443–78.

Sampson, R.J., and Raudenbush, S.W. (1999). Systematic social observation of public spaces: A new look at disorder in urban neighborhoods. *American Journal of Sociology*, 105(3): 603–51.

Sampson, R.J., and Raudenbush, S.W. (2004). Seeing disorder: Neighborhood stigma and the social construction of "Broken windows". *Social Psychology Quarterly*, 67(4): 319–342.

Saunderson, T., and Langford, I. (1996). A Study of the Geographical distribution of suicide rates in England and Wales 1989–1992 using empirical Bayes estimates *Social Science and Medicine*, 43(4): 489–502.

Sawer, M. (1975). Plekhanov and Relationship Between Geographical Determinism And Historical Materialism. *Political Science*, 27(1–2): 117–23.

Saxena, S., Carlson, D., Billington, R., and Orley, J. (on behalf of the WHOQOL Group) (2001). The WHO quality of life assessment instrument (WHOQOL-Bref): The importance of its items for cross-cultural research. *Quality of Life Research*, 10, 711–21.

Saxena, S., Carlson, D., and Billington, R. (2003). The WHO quality of life assessment instrument (WHOQOL-Bref): the importance of its items for cross-cultural research. *Quality of Life Research*, 10, 711–21.

Saxena, S., Sharan, P., Garrido, M., and Saraceno, B. (2006). World Health Organization's Mental Health Atlas 2005: implications for policy development *World Psychiatry*, 5(3): 179–84.

Schama, S. (1995). *Landscape and Memory*. London: HarperCollins.

Schehaye, M. (1970). *Autobiography of a Schizophrenic Girl*. New York: New American Library.

Schell, L.M., and Denham, M. (2003). Environmental pollution in urban environments and human biology. *Annual Review of Anthropology*, 32, 111–34.

Schimmel, P. (2001a). Mind over matter? I: philosophical aspects of the mind-brain problem. *Australian and New Zealand Journal of Psychiatry*, 35(4): 481–7.

Schimmel, P. (2001b). Mind over matter? II: implications for psychiatry. *Australian and New Zealand Journal of Psychiatry*, 35(4): 488–94.

Schneider, S.M. (2007). The dependent gene: The fallacy of "nature vs. nurture". *Behavior Analyst*, 30(1): 91–105.

Schneiders, J., Drukker, M., et al. (2003). Neighbourhood socioeconomic disadvantage and behavioural problems from late childhood into early adolescence *Journal of Epidemiology and Community Health*, 57(9): 699–703.

Schnittker, J., and McLeod, J.D. (2005). The social psychology of health disparities. *Annual Review of Sociology*, 31, 75–103.

Schonberg, M.A., and Shaw, D.S. (2007). Do the predictors of child conduct problems vary by high- and low-levels of socioeconomic and neighborhood risk? *Clinical Child and Family Psychology Review*, 10(2): 101–36.

Schroeder, S.R. (2000). Mental retardation and developmental disabilities influenced by environmental neurotoxic insults. *Environmental Health Perspectives*, 108, 395–99.

Schweitzer, L., and Kierszenbaum, H. (1978). Community characteristics that affect hospitalization and rehospitalization rates in a municipal psychiatric hospital. *Community Mental Health Journal*, 14(1): 63–73.

Scully, V. (1969). *The Earth, the Temple and the Gods: Greek Sacred Architecture*. New York: Praeger.

Seamon, D. (1979). *A Geography of the Lifeworld: Movement, Rest and Encounter*. London: Croom Helm.

Sedgwick, E. (2003). *Touching Feeling: Affect Pedagogy Performativity*. Durham NC: Duke University Press.

Segrott, J., and Doel, M.A. (2004). Disturbing geography: obsessive-compulsive disorder as spatial practice. *Social and Cultural Geography*, 5(4): 597–614.

Self, W. (2007). *Psychogeography*. London: Bloomsbury.

Semple, E.C. (1911). *Influences of the Geographic Environment: On the Basis of Ratzel's System of Anthropo-geography*. New York: Henry Holt.

Sen, A. (1970). *Collective Choice and Social Welfare*. Edinburgh: Oliver and Boyd.

Shama, S. (1996). *Landscape and Memory*. London: Fontana.

Shankar, J., and Muthuswamy, S.S. (2007). Support needs of family caregivers of people who experience mental illness and the role of mental health services. *Families in Society – The Journal of Contemporary Social Services*, 88(2): 302–10.

Shannon, G., and Dever, G. (1974). *Health Care Delivery: Spatial Perspectives*. New York: McGraw Hill.

Shore, C. (1994). Community, in W. Outhwaite and T. Bottomore (eds), *The Blackwell Dictionary of Twentieth Century Social Thought*. Oxford: Blackwells, pp. 98–99.

Sibley, D. (1995). *Geographies of Exclusion: Society and Difference in the West*. London: Routledge.

Siegrist, J., and Marmot, M. (2004). Health inequalities and the psychosocial environment – two scientific challenges. *Social Science and Medicine*, 58(8): 1463–73.

Silman, A. (1995). *Epidemiological Studies: A Practical Guide*. Cambridge: Cambridge University Press.

Silver, E., Mulvey, E.P., and Swanson, J.W. (2002). Neighborhood structural characteristics and mental disorder: Faris and Dunham revisited. *Social Science and Medicine*, 55(8): 1457–70.

Skapinakis, P., Weich, S., Lewis, G., Singleton, N., and Araya, R. (2006). Socio-economic position and common mental disorders – Longitudinal study in the general population in the UK. *British Journal of Psychiatry*, 189, 109–117.

Skelton, T. (2000). 'Nothing to do, nowhere to go?': teenage girls and 'public space in the Rhondda Valleys, South Wales, in S. Holloway, and G. Valentine (eds), *Children's Geographies*. London: Routledge, pp. 80–99.

Smith, B.W., and Freedy, J.R. (2000). Psychosocial resource loss as a mediator of the effects of flood exposure on psychological distress and physical symptoms. *Journal of Traumatic Stress*, 13(2): 349–57.

Smith, C. (1978). Recidivism and community adjustment amongst former mental patients *Social Science and Medicine*, 12, 17–27.

Smith, C., Giggs, J., and Hanham, R. (1981). Any Place But Here – Mental-health Facilities as Noxious Neighbors. *Professional Geographer*, 33(3): 326–34.

Smith, C., and Hanham, R. (1981a). Deinstitutionalization of the Mentally-ill – A Time Path-analysis of the American States, 1955–1975. *Social Science and Medicine Part D: Medical Geography*, 15(3D): 361–78.

Smith, C., and Hanham, R. (1981b). Proximity and the Formation of Public: Attitudes Towards Mental-illness. *Environment and Planning A*, 13(2): 147–65.

Smith, C., and Giggs, J. (eds) (1988). *Location and. Stigma*. Boston: Unwin Hyman.

Smith, D.M. (1977). *Human Geography: A Welfare Approach*. London: Arnold.

Smith, D.M. (1994). *Geography and Social Justice: Social Justice in a Changing World*. London: Wiley.

Smith, J.M. (2008). Identities and Urban Social Spaces in Little Tokyo, Los Angeles: Japanese Americans in Two Ethno-spiritual Communities. *Geografiska Annaler Series B: Human Geography*, 90B(4): 389–408.

Smith, P., Sheldon, T, and Martin, S. (1996). An index of need for psychiatric services based on in-patient utilisation. *British Journal of Psychiatry*, 169(3): 308–16.

Smith, S.J. (2005). States, markets and an ethic of care. *Political Geography*, 24(1): 1–20.

Smyth, F. (2008). Medical geography: understanding health inequalities. *Progress in Human Geography*, 32(1): 119–27.

Song, S., Wang, W., and Hu, P. (2009). Famine, death, and madness: Schizophrenia in early adulthood after prenatal exposure to the Chinese Great Leap Forward Famine. *Soc Sci Med*, 68(7): 1315–21.

Speed, F. (2003). The sacred environment and its implications for place-making, in S.E. Menin (ed.), *Constructing Place: Mind and Matter*. London: Routledge, pp. 55–65.

Spinoza, B. (2000). *Ethics*. Oxford: Oxford University Press.

Stafford, M., Bartley, M., Sacker, A., Marmot, M., Wilkinson, R., Boreham, R., et al. (2003). Measuring the social environment: social cohesion and material deprivation in English and Scottish neighbourhoods. *Environment and Planning A*, 35(8): 1459–75.

Stafford, M., De Silva, M., Stansfeld, S., and Marmot, M. (2008). Neighbourhood social capital and common mental disorder: Testing the link in a general population sample. *Health and Place*, 14(3): 394–405.

Stansfeld, S.A., Berglund, B., Clark, C., Lopez-Barrio, I., Fischer, P., Ohrstrom, E., et al. (2005). Aircraft and road traffic noise and children"s cognition and health: a cross-national study. *Lancet*, 365(9475): 1942–49.

Steinhausen, H., Gundelfinger, R., and Metzke, C. (2009). Prevalence of self-reported seasonal affective disorders and the validity of the seasonal pattern assessment questionnaire in young adults Findings from a Swiss community study. *Journal of Affective Disorders*, 115(3): 347–54.

Stellman, J.M., Smith, R.P., Katz, C.L., Sharma, V., Charney, D.S., Herbert, R., et al. (2008). Enduring mental health morbidity and social function impairment in World Trade Center rescue, recovery, and cleanup workers: The psychological dimension of an environmental health disaster. *Environmental Health Perspectives*, 116(9): 1248–53.

Stephens, C. (2008). Social capital in its place: Using social theory to understand social capital and inequalities in health. *Social Science and Medicine*, 66(5): 1174–84.

Steptoe, A., Tsuda, A., Tanaka, Y., and Wardle, J. (2007). Depressive symptoms, socio-economic background, sense of control, and cultural factors in university students from 23 countries. *International Journal of Behavioral Medicine*, 14(2): 97–107.

Stewart-Brown, S., Tennant, A., Tennant, R., Platt, S., Parkinson, J., and Weich, S. (2009). Internal construct validity of the Warwick–Edinburgh Mental Well-being Scale (WEMWBS): a Rasch analysis using data from the Scottish Health Education Population Survey. *Health and Quality of Life Outcomes*, 7.

Stotz, K. (2006). With 'genes' like that, who needs an environment? Postgenomics's argument for the 'ontogeny of information'. *Philosophy of Science*, 73(5): 905–17.

Sturm, R., and Gresenz, C.R. (2002). Relations of income inequality and family income to chronic medical conditions and mental health disorders: national survey in USA. *British Medical Journal*, 324(7328): 20.

Subramanian, S.V., Jones, K., Kaddour, A., and Krieger, N. (2009a). Response: The value of a historically informed multilevel analysis of Robinson's data. *International Journal of Epidemiology*, 38(2): 370–73.

Subramanian, S.V., Jones, K., Kaddour, A., and Krieger, N. (2009b). Revisiting Robinson: The perils of individualistic and ecologic fallacy. *International Journal of Epidemiology*, 38(2): 342–60.

Subramanian, S.V., Chen, J., Rehkopf, D.H., Waterman, P.D., and Krieger, N. (2005). Racial disparities in context: A multilevel analysis of neighborhood variations in poverty and excess mortality among black populations in Massachusetts. *American Journal of Public Health*, 95(2): 260–5.

Subramanian, S.V., Lochner, K.A., and Kawachi, I. (2003). Neighborhood differences in social capital: a compositional artifact or a contextual construct? *Health and Place*, 9(1): 33–44.

Sugiyama, T., Leslie, E., Giles-Corti, B., and Owen, N. (2009). Associations of neighbourhood greenness with physical and mental health: do walking, social coherence and local social interaction explain the relationships? *Journal of Epidemiology and Community Health*, 62(5): e9 (electronic issue), 1–6.

Summerfield, D. (2001). The invention of post-traumatic stress disorder and the social usefulness of a psychiatric category. *British Medical Journal*, 322(7278): 95–8.

Sustainable Development Commission. (2004). *Outdoor Environments and Health*. The Hague, Netherlands.

Sweeting, H., Young, R., West, P., and Der, G. (2006). Peer victimization and depression in early- mid-adolescence: A longitudinal study. *British Journal of Educational Psychology*, 76, 577–94.

Takahashi, L.M., and Gaber, S.L. (1998). Controversial facility siting in the urban environment – Resident and planner perceptions in the United States. *Environment and Behavior*, 30(2): 184–215.

Tammet, D. (2006). *Born on a Blue Day*. London: Hodder and Stoughton.

Tansella, M., and Thornicroft, G. (1998). A conceptual framework for mental health services: the matrix model. *Psychological Medicine*, 28(3): 503–8.

Tapsell, S., and Tunstall, S. (2007). The mental health aspect of floods: evidence from England and Wales, in R. Few, and F. Matthies (eds), *Flood Hazards and Health*, pp. 89–110.

Tarabrina N., Lazebnaya, E., and Zelenova M. (1994). Psychological features of PTSD of "liquidators": results from the Chernobyl accident [in Russian]. *Psihologicheskiy Jurnal*, 15, 67–77.

Taylor, J., Edwards, J., Kelly, F., and Fielke, K. (2009). Improving transfer of mental health care for rural and remote consumers in South Australia. *Health and Social Care in the Community*, 17(2): 216–24.

Taylor, L., Taske, N., Swann, C., and Waller, S. (2007). Public health interventions to promote positive mental health and prevent mental health disorders among adults: Evidence Briefing. London: National Institute for Clinical Excellence, London, UK.

Tellez-Rojo, M.M., Bellinger, D.C., Arroyo-Quiroz, C., Lamadrid-Figueroa, C., Mercado-Garcia, A., Schnaas-Arrieta, L., et al. (2006). Longitudinal associations between blood lead concentrations lower than 10 mu g/dL and neurobehavioral development in environmentally exposed children in Mexico City. *Pediatrics*, 118(2): E323–E330.

Tennant, R., Hiller, L., Fishwick, R., Platt, S., Joseph, S., Weich, S., et al. (2007). The Warwick–Edinburgh mental well-being scale (WEMWBS): development and UK validation. *Health and Quality of Life Outcomes*, 5.

Tennant, R., Joseph, S., and Stewart-Brown, S. (2007). The Affectometer 2: a measure of positive mental health in UK populations. *Quality of Life Research*, 16(4): 687–95.

Thein, D. (2005). After or beyond feeling? A consideration of affect and emotion in geography. *Area*, 37(4): 450–6.

Thomas, J.L., Jones, G.N., Scarinci, I.C., and Brantley, P.J. (2007). Social support and the association of type 2 diabetes and depressive and anxiety disorders among low-income adults seen in primary care clinics. *Journal of Clinical Psychology in Medical Settings*, 14(4): 351–9.

Thomas, R., Evans, S., Huxley, P., Gately, C., and Rogers, A. (2005). Housing improvement and self-reported mental distress among council estate residents. *Social Science and Medicine*, 60(12): 2773–83.

Thomson, H., Thomas, S., and Sellstrom, E. (2006). Housing improvement and health: a systematic review of world literature (1900–2005). *European Journal of Public Health*, 16, 93.

Thomson, H., Thomas, S., Sellstrom, E., and Petticrew, M. (2008). Best available evidence on housing improvement and health: A systematic review. Journal of Epidemiology and Community Health (p. 011).

Thornicroft, G., Bisoffi, G., De Salvia, D., Tansella, M. (1993). Urban–rural differences in the associations between social deprivation and psychiatric services utiliation in schizophrenia and all diagnoses; a case register study in Northern Italy. *Psychological Medicine*, 23, 487–96.

Thrift, N. (2004). Intensities of feelings: towards a spatial politics of affect. *Geografiska Annaler*, 86(B): 57–78.

Tonnellier, F.O., and Curtis, S. (2005). Medicine, landscapes, symbols: "The country Doctor" by Honore de Balzac. *Health and Place*, 11(4): 313–21.

Tönnies, F. (1887). *Gemeinschaft und Gesellschaft*. Leipzig: Fues's Verlag.

Townley, G., Kloos, B., and Wright, P.A. (2009). Understanding the experience of place: Expanding methods to conceptualize and measure community integration of persons with serious mental illness. *Health & Place*, 15(2): 520–31.

Townsend, P., Phillimore, P., Beattie, A. (1988). *Health and Deprivation: Inequality and the North*. London: Croom Helm.

Treffert, D. (2006). *Extraordinary People: Savant Syndrome*. iUniverse, Incorporated

Trudgill, S. (2008). A requiem for the British flora? Emotional biogeographies and environmental change. *Area*, 40(1): 99–107.

Trudgill, S. (2009). 'You can't resist the sea': evolving attitudes and responses to coastal erosion at Slapton, South Devon. *Geography*, 94, 48–57.

Tuan, Y.-F. (1974). Topophilia: *A Study of Environmental Perceptions, Attitudes and Values*. Englewood Cliffs, NJ: Prentice Hall.

Tunstall, S., Tapsell, S., Green, C., Floyd, P., and George, C. (2006). The health effects of flooding: social research results from England and Wales. *Journal of Water and Health*, 4(3): 365–80.

Twiss, J., Dickinson, J., Duma, S., Kleinman, T., Paulsen, H., and Rilveria, L. (2003). Community gardens: Lessons learned from California healthy cities and communities. *American Journal of Public Health*, 93(9): 1435–38.

Tzoulas, K., Korpela,K, Venn, S. Yli-Pelkonen, V., Kaźmierczak, A.,Niemela, J., James, P. (2007). Promoting ecosystem and human health in urban areas using green infrastructure: a literature review. *Landscape and Urban Planning*, 81(3): 167–78.

Tzoulas, K., Korpela, K., Venn, S., Yli-Pelkonen, V., Kazmierczak, A., Niemela, J., et al. (2007). Promoting ecosystem and human health in urban areas using Green Infrastructure: A literature review. *Landscape and Urban Planning*, 81(3): 167–78.

Uhlhaas, P.J., and Mishara, A.L. (2007). Perceptual anomalies in schizophrenia: Integrating phenomenology and cognitive neuroscience. *Schizophrenia Bulletin*, 33(1): 142–156.

Ulrich, R.S. (1984). View Through a Window May Influence Recovery from Surgery. *Science*, 224(4647): 420–1.

Ulrich, R.S., Simons, R.F., Losito, B.D., Fiorito, E., Miles, M.A., and Zelson, M. (1991). Stress Recovery During Exposure to Natural and Urban Environments. *Journal of Environmental Psychology*, 11(3): 201–30.

Valent, F., Little, D., Bertollini, R., Nemer, L.E., Barbone, F., and Tamburlini, G. (2004). Burden of disease attributable to selected environmental factors and injury among children and adolescents in Europe. *Lancet*, 363(9426): 2032–39.

Van Den Wijngaart, M.A.G., Vernooij-Dassen, M., and Felling, A.J.A. (2007). The influence of stressors, appraisal and personal conditions on the burden of spousal caregivers of persons with dementia. *Aging and Mental Health*, 11(6): 626–36.

Vandemark, L.M. (2007). Promoting the sense of self, place, and belonging in displaced persons: The example of homelessness. *Archives of Psychiatric Nursing*, 21(5): 241–8.

Vandermoere, F. (2008). Psychosocial health of residents exposed to soil pollution in a Flemish neighbourhood. *Social Science and Medicine*, 66(7): 1646–1657.

Vazquez, C., Perez-Sales, P., and Matt, G. (2006). Post-traumatic stress reactions following the March 11, 2004 terrorist attacks in a Madrid community sample: A cautionary note about the measurement of psychological trauma. *Spanish Journal of Psychology*, 9(1): 61–74.

Veenstra, G. (2005). Location, location, location: contextual and compositional health effects of social capital in British Columbia, Canada. *Social Science and Medicine*, 60(9): 2059–71.

Veil, S., Salim, E., Izmerov, N.F., Cheng, C.M., Doll, R., Horwitz, A., et al. (1992). Health and the Environment – A Global Challenge. *Bulletin of the World Health Organization*, 70(4): 409–13.

Veling, W., Susser, E., van Os, J., Mackenbach, J.P., Selten, J.P., and Hoek, H.W. (2008). Ethnic density of Neighborhoods and incidence of psychotic disorders among immigrants. *American Journal of Psychiatry*, 165(1): 66–73.

Vine, D.J. (1993). Reflection and revelation: knowing land, places and ourselves, in J.E. Swan (ed.), *Power of Place: Sacred Ground in Natural and Human Environments*. London: Gateway, pp. 28–40.

Vohs, K.D., and Schooler, J.W. (2008). The value of believing in free will – Encouraging a belief in determinism increases cheating. *Psychological Science*, 19(1): 49–54.

Wainer, J., and Chesters, J. (2000). Rural mental health: neither romanticism nor despair. *Aust J Rural Health*, 8(3): 141–7.

Wakefield, J. (2009). Multi-level modelling, the ecologic fallacy, and hybrid study designs. *International Journal of Epidemiology*, 38(2): 330–6.

Wakefield, S., and McMullan, C. (2005). Healing in places of decline: (re)imagining everyday landscapes in Hamilton, Ontario. *Health and Place*, 11(4): 299–312.

Waldrop, M. (1993). *The Emerging Science at the Edge of Order and Chaos*. London: Pocket Books.

Waller, S., and Finn, H. (2004). *Enhancing the Healing Environment: A Guide for NHS Trusts*. London: King's Fund.

Wallerstein, N., and Bernstein, E. (1988). Empowerment Education – Freire Ideas Adapted to Health-education. *Health Education Quarterly*, 15(4): 379–94.

Wang, H.M., Schlesinger, M., Wang, H., and Hsiao, W.C. (2009). The flip-side of social capital: The distinctive influences of trust and mistrust on health in rural China. *Social Science and Medicine*, 68(1): 133–42.

Ware, J., and Sherbourne, C. (1992). The MOS 36-Item Short-Form Health Survey (SF–36®): I. conceptual framework and item selection. *Medical Care*, 30(6): 473–83.

Ware J., and Sherbourne, C. (2009). Sf36 Health Survey Update. Retrieved 30.06.09, from http://www.sf-36.org/tools/sf36.shtml.

Warfa, N., Bhui, K., Craig, T., Curtis, S., Mohamud, S., Stansfeld, S., McCrone, P., Thornicroft, G. (2006). Post-migration geographical mobility, mental health and health service utilisation among Somali refugees in the UK: A qualitative study. *Health and Place*, 12(4): 503–15.

Weber, E. (2006). Experience-based and description-based perceptions of long-term risk: why global warming does not scare us (yet). *Climatic Change*, 77, 103–120.

Weems, C.F., and Overstreet, S. (2008). Child and adolescent mental health research in the context of Hurricane Katrina: An ecological needs-based perspective and introduction to the special section. *Journal of Clinical Child and Adolescent Psychology*, 37(3): 487–94.

Weich, S., Blanchard, M., Prince, M., Burton, E., Erens, B., and Sproston, K. (2002). Mental health and the built environment: cross-sectional survey of individual and contextual risk factors for depression. *British Journal of Psychiatry*, 180, 428–33.

Weich, S., Churchill, R., Lewis, G., and Mann, A. (1997). Do socio-economic risk factors predict the incidence and maintenance of psychiatric disorder in primary care? *Psychological Medicine*, 27(1): 73–80.

Weich, S., and Lewis, G. (1998). Material standard of living, social class, and the prevalence of the common mental disorders in Great Britain. *Journal of Epidemiology and Community Health*, 52(1): 8–14.

Weich, S., Lewis, G., and Jenkins, S P. (2001). Income inequality and the prevalence of common mental disorders in Britain. *British Journal of Psychiatry*, 178, 222–7.

Weich, S., Twigg, L., and Lewis, G. (2006). Rural/non-rural differences in rates of common mental disorders in Britain – Prospective multilevel cohort study. *British Journal of Psychiatry*, 188, 51–7.

Weiss, D., and Marmar, C. (1997). The Impact of Event Scale – Revised, in J. Wilson and T. Keane (eds), *Assessing Psychological Trauma and PTSD*. New York: Guilford Press.

Welch, R.V., and Panelli, R. (2007). Questioning community as a collective antidote to fear: Jean-Luc Nancy's 'singularity' and 'being singular plural'. *Area*, 39(3): 349–56.

West, P., and Sweeting, H. (2003). Fifteen, female and stressed: changing patterns of psychological distress over time. *Journal of Child Psychology and Psychiatry and Allied Disciplines*, 44(3): 399–411.

Western, J., Stimson, R., Baum, S., and Van Gellecum, Y. (2005). Measuring community strength and social capital. *Regional Studies*, 39(8): 1095–1109.

Wheaton, B., and Clarke, P. (2003). Space meets time: Integrating temporal and contextual influences on mental health in early adulthood. *American Sociological Review*, 68(5): 680–706.

Whitebead, H. (2007). Learning, climate and the evolution of cultural capacity. *Journal of Theoretical Biology*, 245(2): 341–50.

Whitehead, M. (1995). *Tackling Inequalities: A Review of Policy Initiatives*. London: King's Fund Publishing.

Whiteley, S. (2004). The evolution of the therapeutic community. *Psychiatric Quarterly*, 75(3): 233–48.

Whitley, E., Gunnell, D., Dorling, D., et al. (1999). Ecological fragmentation, poverty and suicide. *British Medical Journal*, 319, 1034–037.

Whitley, R., Prince, M., McKenzie, K., and Stewart, R. (2006). Exploring the ethnic density effect: A qualitative study of a London electoral ward. *International Journal of Social Psychiatry*, 52(4): 376–91.

WHO (1992). *Psychosocial Consequences of Disasters: Prevention and Management*. Geneva: World Health Organization.

WHO (2001). *The World Health Report 2001 – Mental Health: New Understanding, New Hope*. Geneva: World Health Organisation.

WHO (2005). *Mental Health Atlas 2005*. Geneva: World Health Organization.

WHO (2005). *Mental Health Atlas. 2005*. Revised edition. Geneva: World Health Organization.

WHO European Centre for Health Policy. (1999). Health impact assessment: main concepts and suggested approach. Gothenburg consensus paper. Brussels: WHO Regional Office for Europe, ECHP.

WHO Regional Office for Europe. (2005). Mental Health: Facing the Challenges, Building Solutions. Report from the WHO Ministerial Conference. Copenhagen: WHO European Region.

WHOQOL Group (1998). The World Health Organization quality of life assessment (WHOQOL): Development and general psychometric properties. *Social Science and Medicine*, 46(22): 1569–85.

WHOQOL SRPB Group A cross-cultural study of spirituality, relion and personal beliefs as components of quality of life. *Social Science and Medicine*, 62, 1486–97.

Wickrama, K., and Bryant, C. (2003). Community context of social resources and adolescent mental health. *Journal of Marriage and Family*, 65, 850–66.

Wickrama, K.A.S., Conger, R.D., Lorenz, F.O., and Jung, T. (2008). Family Antecedents and Consequences of Trajectories of Depressive Symptoms from Adolescence to Young Adulthood: A Life Course Investigation. *Journal of Health and Social Behavior*, 49(4): 468–83.

Wickrama, K.A.S., and Kaspar, V. (2007). Family context of mental health risk in Tsunami-exposed adolescents: Findings from a pilot study in Sri Lanka. *Social Science and Medicine*, 64(3): 713–23.

Wickrama, K.A.S., Merten, M.J., and Elder, G.H. (2005). Community influence on precocious transitions to adulthood: Racial differences and mental health consequences. *Journal of Community Psychology*, 33(6): 639–63.

Wiersma, E.C. (2008). The experiences of place: Veterans with dementia making meaning of their environments. *Health and Place*, 14(4): 779–94.

Wildman, J. (2003). Income related inequalities in mental health in Great Britain: analysing the causes of health inequality over time. *Journal of Health Economics*, 22(2): 295–312.

Wilkinson, R. (1996). *Unhealthy Societies: The Afflications of Inequality*. London: Routledge.

Wilkinson, R., Pickett, K. (2009). *The Spirit Level: Why More Equal Societies Almost Always Do Better*. London: Allen Lane.

Wilkinson, R.G., and Pickett, K.E. (2007). The problems of relative deprivation: Why some societies do better than others. *Social Science and Medicine*, 65, 1965–78.

Williams, A. (ed.) (1998). *Therapeutic Landscapes: The Dynamic between Place and Wellness*. Boston: University Press.

Williams, A. (ed.) (2007). *Therapeutic Landscapes*. Aldershot: Ashgate.

Williams, S.J. (2000). Reason, emotion and embodiment: is 'mental' health a contradiction in terms? *Sociology of Health and Illness*, 22(5): 559–81.

Wilson, E.O. (1984). *Biophilia*. Cambridge, MA: Harvard University Press.

Wilson, K. (2003). Therapeutic landscapes and First Nations peoples: an exploration of culture, health and place. *Health and Place*, 9, 83–93.

Wilson, M., Robertson, L., Davly, M., and Walton, S. (1995). Effects of visual cues on assessment of water quality. *Journal of Environmental Psychology*, 15, 53–63.

Wiltshire, S. (1989). *Cities*. London: Dent and Son.

Wing, S. (2003). Objectivity and ethics in environmental health science. *Environmental Health Perspectives*, 111(14): 1809–18.

Wolch, J., and Philo, C. (2000). From distributions of deviance to definitions of difference: past and future mental health geographies *Health and Place*, 6(3): 137–57.

Wolford, W. (2005). Political ecology: A critical introduction. *Annals of the Association of American Geographers*, 95(3): 717–19.

Wood, L., and Giles-Corti, B. (2008). Is there a place for social capital in the psychology of health and place? *Journal of Environmental Psychology*, 28(2): 154–63.

Xue, Y.G., Leventhal, T., Brooks-Gunn, J., and Earls, F.J. (2005). Neighborhood residence and mental health problems of 5- to 11-year-olds. *Archives of General Psychiatry*, 62(5): 554–63.

Yantzi, N.M., and Rosenberg, M.W. (2008). The contested meanings of home for women caring for children with long-term care needs in Ontario, Canada. *Gender Place and Culture*, 15(3): 301–15.

Yen, I.H., and Kaplan, G.A. (1999). Poverty area residence and changes in depression and perceived health status: evidence from the Alameda County Study. *International Journal of Epidemiology*, 28(1): 90–4.

Yeung, H.W.C. (2005). Rethinking relational economic geography. *Transactions of the Institute of British Geographers*, 30(1): 37–51.

Yevelson, II, Abdelgani, A., Cwikel, J., and Yevelson, I.S. (1997). Bridging the gap in mental health approaches between east and west: The psychosocial consequences of radiation exposure. *Environmental Health Perspectives*, 105, 1551–56.

Yip, W., Subramanian, S.V., Mitchell, A.D., Lee, D.T.S., Wang, J., and Kawachi, I. (2007). Does social capital enhance health and well-being? Evidence from rural China. *Soc Sci Med*, 64(1): 35–49.

Young, A. (1995). *The Harmony of Illusion: Inventing Post-Traumatic Stress Disorder*. New Jersey: Princeton.

Young, M., and Willmott, P. (1957). *Family and Kinship in East London*. London: Institute of Community Studies [Pelican, 1962].

Yuan, A.S.V. (2008). Racial composition of neighborhood and emotional well-being. *Sociological Spectrum*, 28(1):105–29.

Zechmeister, I., and Osterle, A. (2007). Informal care of people with mental disorders: Does the Austrian long-term care system provide adequate support? *Neuropsychiatrie*, 21(1): 29–36.

Zimmerman, F.J., and Bell, J.F. (2006). Income inequality and physical and mental health: testing associations consistent with proposed causal pathways. *Journal of Epidemiology and Community Health*, 60(6): 513–21.

Index

Figures are indicated by **bold** page numbers